DATA SCIENCE
FOR
IoT ENGINEERS

LICENSE, DISCLAIMER OF LIABILITY, AND LIMITED WARRANTY

DATA SCIENCE

FOR

IoT ENGINEERS

A Systems Analytics Approach

P. G. Madhavan, Ph.D.

MERCURY LEARNING AND INFORMATION
Dulles, Virginia
Boston, Massachusetts
New Delhi

Publisher: David Pallai
MERCURY LEARNING AND INFORMATION
22841 Quicksilver Drive
Dulles, VA 20166
info@merclearning.com
www.merclearning.com
1-800-232-0223

P. G. Madhavan. *Data Science for IoT Engineers*.
ISBN: 978-1-68392-642-9

The publisher recognizes and respects all marks used by companies, manufacturers, and developers as a means to distinguish their products. All brand names and product names mentioned in this book are trademarks or service marks of their respective companies. Any omission or misuse (of any kind) of service marks or trademarks, etc. is not an attempt to infringe on the property of others.

Library of Congress Control Number: 2021942159
212223321   Printed on acid-free paper in the United States of America.

Our titles are available for adoption, license, or bulk purchase by institutions, corporations, etc. For additional information, please contact the Customer Service Dept. at 800-232-0223(toll free).

All of our titles are available in digital format at *academiccourseware.com* and other digital vendors. The sole obligation of MERCURY LEARNING AND INFORMATION to the purchaser is to replace the book, based on defective materials or faulty workmanship, but not based on the operation or functionality of the product.

To my wife, Ann

CONTENTS

PREFACE

This book is the third iteration of the book I originally published in 2016 as "Systems Analytics." The title reflected a new development effort in the field of machine learning, grounded firmly in systems theory. My intention in writing the first edition was to bring mathematically trained graduates in engineering, physics, mathematics, and allied fields into data science.

The objective of this edition, *Data Science for IoT Engineers*, remains the same. Part I, where I develop machine learning (ML) algorithms from the background of engineering courses such as control theory, signal processing, etc. is largely unchanged. However, dynamical systems-based Part II now takes a more detailed Multi-Input-Multi-Output (MIMO) systems approach, and develops a new and important form of digital twins called *"causal" digital twin*. This topic is significant, because on the one hand, *digital twin is the seat of ML & AI in IoT solutions, but more important, causality is a critical factor in enabling "prescriptive analytics," which is the real promise of Internet of Things (IoT)*. An epilogue has been added that introduces a new theory of random fields; it is shown that a new second-order property might have significant practical applications, and some of these are discussed.

PART I MACHINE LEARNING FROM MULTIPLE PERSPECTIVES

In this part, we bring together machine learning, systems theory, linear algebra and digital signal processing. The intention is to make clear the similarity of basic theory and algorithms among these disparate fields. Hands-on exposure to machine learning is provided. This part concludes with a complete description of modern machine learning and a new ontology grounded in probability theory.

PART II SYSTEMS ANALYTICS

With the realization that business solutions are not "one and done" and they require ongoing measurement, tracking and fine-tuning, we embed machine learning in a closed-loop, real-time systems framework – adaptive machine learning This naturally leads to the formal development of state-space formulation, Bayesian estimation, and Kalman filter. We develop a "universal" nonlinear, time-varying, dynamical machine learning solution which can faithfully model all the essential complexities of real-life business problems, and show how to apply it.

Developing the systems theme into the framework of digital twins as the action center for IoT-related machine learning, we explore several types of digital twins: (1) display, (2) forward and (3) inverse. *Inverse digital twin* is an example of a powerful form of "causal" modeling that captures the "dynamics" or "kinetics" of machinery in industrial applications. In the epilogue, we introduce some future development possibilities for data science from a "complexity" point of view.

Intended Audience

This book is neither a hard-core university text nor a popular science read. STEM enthusiasts of all ilk are introduced to data science in ways that leverage their engineering *science* background. Newly minted data scientists can see the larger framework beyond the bag-of-tricks that they learned in their coursework and glimpse the future of "adaptive" machine learning (what we call "systems analytics"). In addition, one can expect to clear the confusion of multiplying digital twin names and claims, while developing a full understanding of the "next-gen" causal digital twins.

P. G. Madhavan, Ph.D.
October 2021

About the Author

P. G. Madhavan, Ph.D. has an extensive background in IoT, machine learning, digital twin, and wireless technologies in roles such as Chief IoT Officer, Chief Acceleration Officer, IoT Startup Founder, IoT Product Manager at large corporations (Rockwell Automation, GE Aviation, NEC), and small firms and startups. After obtaining his Ph.D. in electrical and computer engineering from McMaster University, Canada, and a Master's degree in biomedical engineering from IIT, Madras, Dr. Madhavan pursued original research in random field theory and computational neuroscience as a professor at the University of Michigan, Ann Arbor and Waterloo University, Canada, among others. His next career in corporate technology saw him assume product leadership roles at Microsoft, Bell Labs, Rockwell Automation, GE Aviation, and lastly at NEC. He has founded and was CEO at two startups (and CTO at two others) leading all aspects of startup life. Currently, he champions digital twins as the seat of AI/ML for IoT applications with an emphasis on causality.

MACHINE LEARNING FROM MULTIPLE PERSPECTIVES

OVERVIEW OF DATA SCIENCE

This chapter includes insight into machine learning, data science, systems theory, digital twins, and artificial intelligence (AI), as well as their business relevance to the reader. In the following chapters, we will discuss the models and methods of these topics more formally. Machine learning (ML) using Big Data generates business value by predicting what profitable actions to take and when.

Successful businesses find a way to understand and manage complexity. Managing complex systems requires effectively using a significant amount of data at the right time. Big Data provides the data we need. To put the data to work, we have to anticipate what is about to happen and react when it happens in a closed loop manner. Predictive analytics allows us to push our system to the edge (without "falling over") in a managed fashion. Now, an increase in prediction accuracy from 82% to 83% may not seem like much, but the business effects of that 1% improvement can be disproportionate (for example, a consumer conversion rate of 2% can jump to 20%). Businesses embrace predictive analytics to manage their business at a high level of performance and achieve excellent business results at the edge of complexity overload [CJ13].

Machine learning began in earnest in 1973 with the publication of Duda and Hart's classic textbook *Pattern Classification & Scene Analysis* [DR73]. In the early days when computer processing power and memory were severely limited, the focus was to use as little data as possible to extract as much information as possible. Admittedly, the results were not very good and data specialists were not involved in the process. The advent of Big Data and the strong focus on developing algorithms that extract as much information as possible means that data analysts are an important part of business success.

One of the best methods to minimize the amount of data needed was to use each data point as it arrived, without storing all the past data and re-doing all the calculations. The past result was updated based on the new data point. This is what we referred to as "adaptive or recursive" methods. This process is also known as "learning." Another early approach was to try to understand and replicate learning in living organisms via perceptrons and cybernetics (which encompasses the control and communication between an animal and a machine). These factors led to algorithmic learning, which is the technical basis of today's popular technology, machine learning.

What do we need to experience success with data? To start with, we need both machine learning and analytics. We also require pattern recognition, statistical modeling, predictive analytics, data science, adaptive systems, and self-organizing systems. We won't worry too much about nuanced meanings (there are differences, but not in our present context).

Note AI is not included in this list. Creating AI by mimicking the human brain seems to be a fool's errand. If you think about the neuronal axons along which electrical spikes travel, they are like billions of conducting wires with the insulation scraped off every few millimeters. These billions of wires are also in a salt solution. Try sending your TCP/IP packets along such a network.

A better field to pursue is that of Intelligence Augmentation(IA). When Doug Englebart wrote to ARPA about IA in the 1960s, he could not have foreseen what Big Data could do. Englebart's "Mother of All Demos" was really about *communication* augmentation using a computer mouse, video conferencing, and hypertext. With Big Data and analytics, we can truly perform intelligence augmentation.

Prediction is the foundational requirement in ML business use cases. Let us explore in some detail what prediction entails. Wanting to know the future has always been a human preoccupation. You cannot truly know the future, but in some cases, predictions are possible.

We should consider short-term and long-term predictions separately. Long-term prediction is nearly impossible. In the 1980s and 1990s, chaos and complexity theorists showed us that things can become uncontrollable even when we have perfect past and present information (for example, predicting the weather beyond three weeks is a major challenge, if not impossible). Stochastic process theory tells us that "non-stationarity," where statistics evolve (slowly or fast), can render longer term predictions unreliable.

If the underlying systems do not evolve quickly or suddenly, there is some hope. Causal systems (in systems theory, no future information of any kind is available in the current state of the system) indicate that outcomes are predictable in the sense that, as long as certain conditions are met, we can be somewhat confident in predicting a few steps ahead. This may be quite useful in some data science applications (such as in fintech).

Another type of prediction involves not the actual path of future events (or the "state space trajectories"), but the occurrence of a "black swan" or an "X-event" (for an elegant in-depth discussion, see [CJ13]). Any unwanted event can be good to know about in advance. Consider unwanted destructive vibrations (called "chatter") in machine tools, as an example; early warning may be possible and very useful in saving expensive work pieces [MP97]. We find that sometimes the underlying system does undergo some pre-event changes (such as approaching the complexity overload, state-space volume inflation, and increase in degrees of freedom) which may be detectable and trackable. However, there is no way to prevent false positives (and the associated waste of resources preparing for an event that does not come) or false negatives (and be blind-sided when we are told it is not going to happen).

We will use an explicit systems theory approach to analytics. In our system analytics formulation, the parameters of the system and its variation over time are tracked adaptively in real time, and so it can tell us how long into the future we can predict safely. If the parameters evolve slowly or cyclically, we can have a higher confidence in our predictive analytics solutions.

Machine learning has to do with learning, i.e., the ability to generalize from experience. A necessary feature of learning is feedback, either explicit, as in the case of supervised learning, or implicit, as in the cases of unsupervised or reinforcement learning.

There is a class of algorithms that is very useful but does not exhibit the learning and feedback behaviors. We will exclude them from our current discussion of analytics and ML. Algorithms such as decision trees are more associated with data mining than ML. Such approaches are useful but do not involve learning or feedback about the information itself.

We will select a business context to explore ML. There is rarely a business solution that requires a one-time answer that will completely solve the problem. Almost all business problems you'll typically encounter require some type of regular attention. Analysts need to monitor the outcomes of the first solution, tweak the approach, and apply it again after a while.

Data scientists have the duty to educate business clients to subscribe to this view, what we call "goal-seeking" or the tracking solution concept. The first solution may just provide an 80% solution to the original problem, but with tracking, it can improve over time. With such realistic expectations, your customer will be delighted if the trajectory of improvement is good and fast.

With this preamble, let us define a canonical or prototypical business problem that will help crystallize our ML approach and the analytics roadmap ahead of us.

CANONICAL BUSINESS PROBLEM

Let us consider a retail commerce business, a brick-and-mortar grocery store chain, and the CPG (consumer product goods) manufacturers who supply them with fast-moving consumer goods (FMCG).

Consider a grocery retail chain with 100 stores. Assume that each store has 5 departments, 20 product categories per department, 50 brands per category, and 70 SKUs per brand. The product category is the level in the product hierarchy to focus on, since shoppers' choices are within a category. Each store then has 100 categories and 1,000 brands and 70,000 SKUs.

Consider a neighborhood and stores in the vicinity; the activities near each store can be dramatically different (for example, one may have a school nearby, and another may have offices near it). How does the store manager at one of these stores decide what products to carry? This is called the Optimal Product Assortment problem in retail merchandising. This is important not only to the retailer, but also the CPGs that supply products to the store.

The store manager has several possible options: they can make (1) educated guesses (not very reliable), (2) ask the shoppers what they want (does not scale), and (3) rely on market survey data (this is usually based on sampling and is not specific to SKU and store). Using Big Data and analytics, can we do better?

A BASIC ML SOLUTION

Recommendation engines are commonplace today – such as Amazon or Netflix. They provide recommendations to an individual shopper (for goods or movies) based on a particular shopper's likes and dislikes as well as the sellers' business priorities (push out movies in the "long tail" for Netflix and sell goods in the "fat front" from Amazon warehouses).

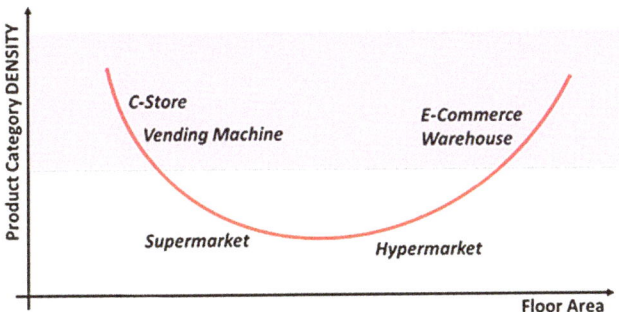

Figure 1.1 Product density in stores

Optimal product assortment is a slightly more difficult problem. We need a recommendation engine for a group of shoppers, one for each store. Note that the optimization problem arises because shelf space is limited; this constraint applies equally to large ecommerce warehouses where the shelves are numerous, but the number of products to stock is significant. The challenge is the product density, which varies, as shown in Figure 1.1.

The natural approach to such a problem is to group the hundreds or thousands of shoppers at a store via segmentation or clustering. Once grouped into a few segments, the recommendation engine can be optimized for each segment, and the optimum product assortment can be derived from the proportion of these few segments that shop at the store.

The state-of-the-art in commerce is *behavioral segmentation*, where the market is divided into segments based on pre-selected characteristics that apply to all product categories in a store.

Clearly, placing shoppers into convenient segments such as "Price sensitive" or "Families," as shown in Figure 1.2, allows one level of meaningful abstraction. Instead of addressing millions of shoppers individually, one can tailor marketing, merchandising, and loyalty efforts to a handful of labelled groups.

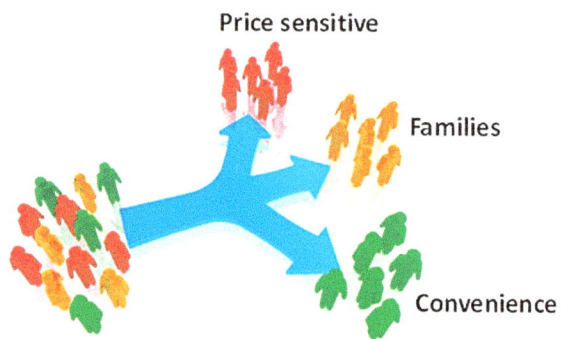

Figure 1.2 Behavioral segmentation example

However, what is helpful at one level can be a flawed approach for some applications. Consider a case where a particular shopper, per behavioral segmentation, ended up in the "Price sensitive" group. While this may be true in general for them, they may have specific preferences in certain product categories; for example, while "Price sensitive" in general, their wine choice may be the expensive Châteauneuf-du-Pape brand. Such misallocations, when multiplied by millions of shoppers, lead to flawed product assortment decisions when behavioral segmentation is applied to merchandising. Let us consider a better approach using shopper data and machine learning to create and identify segments.

In ML segmentation, the shoppers (whatever their behavioral characteristics may be) fall into N Preference Groups (Figure 1.3) based on what they actually buy (actual purchase patterns are a great proxy for true product preference). In essence, each product category is its own unique market.

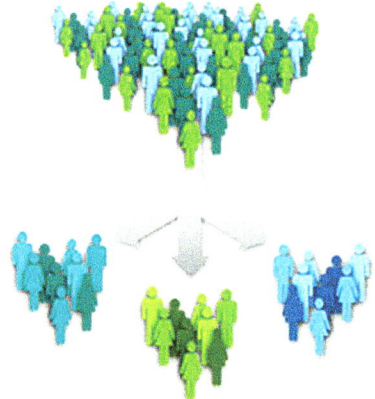

Figure 1.3 Preference groups

Traditional behavioral segmentation would have predicted that Brand X will sell more in Store 123 because "Price sensitive" shoppers prefer more of Brand X and Store 123 has more "Price sensitive" shoppers.

In the ML method, we realize that since Store 123 shoppers are well-represented by N preference groups for a particular product category, the proportion of the N groups that shop at Store 123 determines the assortment for that product category at Store 123. Making such fine distinctions with the aid of shopper data avoids the pitfall of employing the same behavioral groups across all product categories because shoppers' purchase propensities can vary across categories.

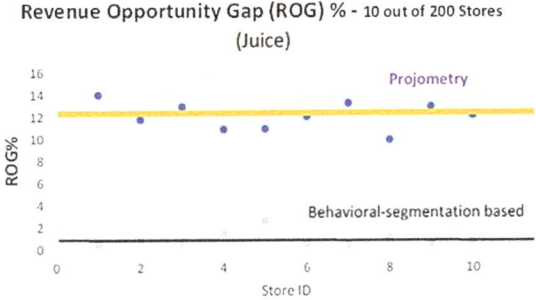

Figure 1.4 Revenue opportunity gap

By comparing behavioral segmentation and the ML method to optimize product assortments, we obtained the results shown in Figure 1.4. Consider the *Revenue Opportunity Gap* (ROG) as an overall performance measure, which indicates a better product assortment optimization when the value is high. Assume that the overall revenue for a category (yogurt, in this example) was $100. While behavioral segmentation shows an average of 1% or $1 of ROG (improvement possibility), Projometry™, which is an ML segmentation method, shows a 5% or $5 improvement possibility. In other words, the improvement due to our data-driven method is five times higher than that due to behavioral segmentation.

Why is the ML-based approach better than behavioral segmentation? Table 1.1 shows the reasons. Looking at more dimensions than just the head-to-head outcome comparison above, it is clear that the ML method has several advantages when it comes to merchandising. Whenever the data determine groups rather than being externally imposed, the results will be superior to those of other methods. Another nice feature is that human labor and subjectivity in the segmentation process can be avoided, which makes analysis fast, inexpensive, and repeatable. The fact that separate preference groups are generated for every product category and that the ML method acts on what people purchase rather than why has led to breakout applications of this ML method in retail merchandising product assortment optimization.

Table 1.1 Differences between segmentation approaches

Behavioral Segmentation	ML Method
Simplified model of demand, pre-determined number of segments	Machine determined model of demand
Sample based	Population based
Same across all product categories	Differs by product category
Subjective (behavioral assumption)	Objective (no behavioral assumptions)
Costly and time consuming to create and maintain	Cheap, fast. No segmentation required
Based on WHY people purchase	Focused on WHAT people purchase

In developing this group recommendation engine, you should have noticed that we have added an aspect of "modeling" to the solution. We modelled the shoppers into N preference groups. Alternatively, it can be said that any shopper can be approximated as a weighted linear combination of these N Groups. Let us keep this perspective in mind for systems analytics, which explicitly uses model-based thinking and algorithms.

SYSTEMS ANALYTICS

Remember the concept of "goal-seeking and tracking solutions?" This is the context of systems analytics. The Product Assortment Optimization problem and solution can be reformulated as the following, keeping in mind that we have to track the solution over time. In the retail case, the shopper preferences change, product attributes change, and the local population around a store changes, some quickly and some slowly over time. We want to adapt to the changing circumstances and provide frequent updates of the product assortment solution to the store manager.

A complete characterization of retail dynamics is captured in the canonical diagram in Figure 1.5. In systems theory, this is called a *MIMO system* (Multi-Input Multi-Output system). This basic framework informs all of the development described in this book.

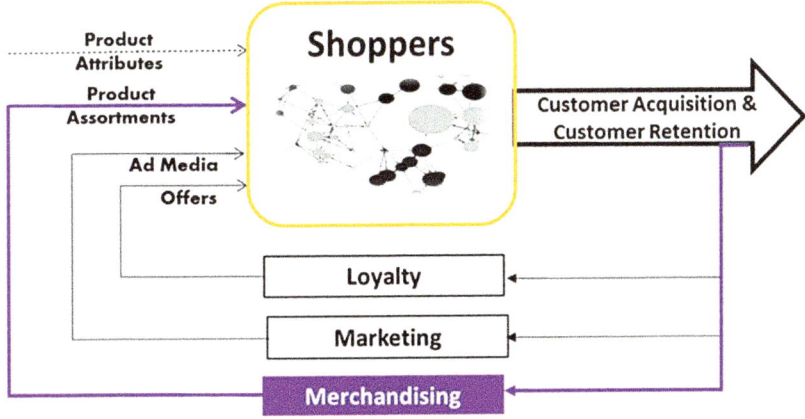

Figure 1.5 Systems analytics model

Retail business is about increasing customer acquisition and retention. Business owners have three areas to affect change: marketing, loyalty, and merchandising. Our prototypical business problem laid out earlier is in retail merchandising and so we will continue to work with that example.

For our basic solution, instead of behavioral segmentation, we used ML to discover N preference groups (or models) to create a group recommendation engine for each store.

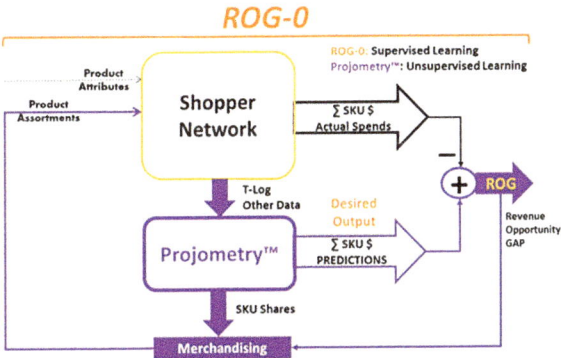

Figure 1.6 Group recommendation engine

Here is a specific instance of such a solution from Syzen Analytics (Figure 1.6) where they use an ML algorithm called Projometry™ for the preference groups (using unsupervised learning) and create optimal product assortment for each store (SKU shares output). The overall group recommendation engine for each store generates feedback information (ROG) for supervised learning, which tracks the optimal product assortments over many quarters.

From this example, we can formalize ML within the systems theory context of systems analytics.

Notice that the following is the essential structure of an ML system, which is shown in Figure 1.7. Using training examples, the learning algorithm tries to "learn" the target function, f, to obtain a good approximation, g, using a model or formula.

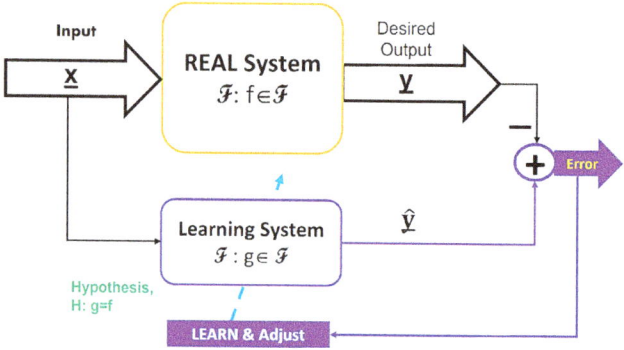

Figure 1.7 System theoretic machine learning

To be explicit:

- We have a set of models.

- Each model in the set has a model structure and an associated learning algorithm.

- Pick a model structure and associated learning algorithm. From the infinite set of functions, F, possible within these two choices, there is the true target function, f, which is an unknowable, and its good approximation, g, that we try to find using our choice of the model structure and associated learning algorithm and training data.

There are many well-known models, structures, and learning algorithms, some of which are shown in Table 1.2.

Table 1.2 Machine learning models

Model	Structure	Learning Algorithm	Objective
Perception	Perception nodes	Perception algorithm	
Logistic regression	Multiple regression	Gradient descent	Find g ≈ f
Neural network	Multiplayer perception	Back propagation	
Systems	*State-space*	*Kalman Filter*	

The state-space model structure and Kalman Filter algorithm define the basic systems analytics framework. There are many simplifications possible; for example, the Markov model is a special case of state-space model, typically used when there is context sensitive information in the input (such as recognizing a word in a sentence). Instead of the Kalman Filter, there are other linear algorithms, such as the Least Mean Square, and non-linear ones, such as the extended Kalman Filter (used in GPS).

The power of the systems analytics formulation is that many other models and algorithms that are well developed in systems theory, control theory, process control, systems identification, and the stochastic process can be brought to bear on ML problems and perhaps lead to powerful solutions that have a formal basis in mathematics (thus, they go beyond ad hoc and heuristic approaches).

From this brief discussion of a systems analytics business example: we can see how systems analytics is equivalent to the ML framework. We have a framework for a closed-loop analytics solution, and a framework for real-time analytics. We also obtained a goal-seeking solution that improves and delivers results over time. Time period may be weeks or days or milliseconds, depending on the application.

DIGITAL TWINS

In this section, we take a preliminary look at a framework by which humans can interact with IoT system analytics solutions in a natural way. There is general acceptance that IoT is a "must have" technology for industry, enterprises, and homes. What is less clear to buyers and sellers of IoT is how to get business value out of IoT deployments. Let us agree that the way businesses get value from the current AI and machine learning boom is through prescriptive analytics. We use it here to mean software that will tell business leaders what to do and when to do it.

In 2000, NASA began the Numerical Propulsion System Simulation or "NPSS." Its objective was stated as follows: "The analysis is currently focused on large-scale modeling of complete aircraft engines. This will provide the product developer with a "virtual wind tunnel" that will reduce the number of hardware builds and tests required during the development of advanced aerospace propulsion systems." (Lytle, John. (2000). The Numerical Propulsion System Simulation: An Overview.) In subsequent years, every jet engine manufacturer (such as GE and Rolls Royce) and many other independent parties (research institutes and universities) became part of this open source effort. Now the NPSS ("gas path") model of a jet engine is used by manufacturers for ML-based fault detection by proprietary methods, which has extended NPSS to suit their "gas path" combined with deep-learning. NPSS is one type of digital twin.

In simple terms, a digital twin is a software counterpart of a physical twin. Here are some examples.

- A physical twin could be a motor, a CNC machine, and a flexible manufacturing system (FMS).

- Physical twin examples can be as varied as a widget or a patient or a whole city.

A digital twin can take many forms. It could be a simple dash-board that displays real-time and historical data from the physical twin or an NPSS-like simulation of a jet engine color-coded for heat buildup and a smart city simulation.

Figure 1.8 shows there are different varieties of digital twins (DTs).

At the outset, we should note that each of the digital twins in Figure 1.8 are not distinct entities, but each are built upon the previous ones.

Display DT is a software image that is populated by data generated from its physical twin (PT). Just this simple level of DT can be very useful in an IoT solution. The human eye is very good at detecting patterns and Display DT can quickly tell if its PT operating properly.

Forward DT is built on a software simulation of a PT. Consider a physical printed circuit board (PCB) of an electronic device. The thermal

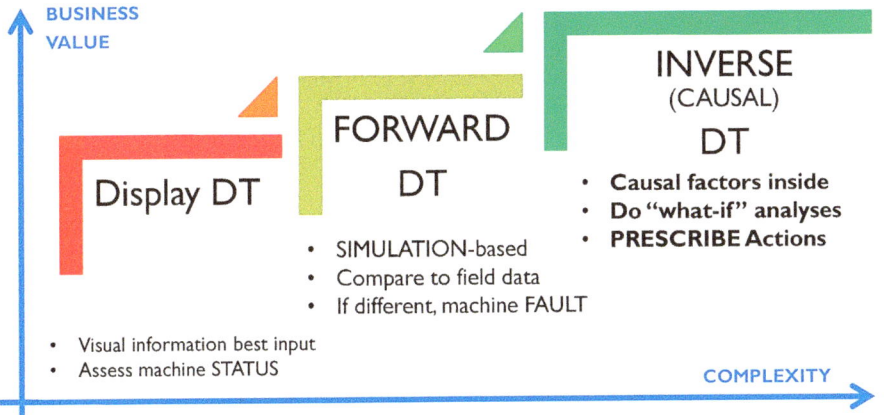

Figure 1.8 Types of digital twins

distribution across the PCB is usually of great interest (consider if you need a fan in the enclosure to cool the PCB for proper operation). Heat diffusion and radiation equations are well-developed, and given the geometry of the PCB, an accurate heat map of the PCB can be calculated on the computer. This heat map can be displayed on a screen to set the baseline; when the electronic device is in operation, the actual measured temperature(s) can be overlaid on the calculated heat map. Any difference in patterns and colors indicate a malfunctioning PCB (or wrong simulation calculations).

Inverse DT is a new and a necessary form of digital twins. We will discuss this topic in great detail in Part II of this book. In addition to the information that Display and Forward DTs provide, we want to know the causes of what we see on the DT in real-time. In that sense, Inverse DT can also be called a causal digital twin. The word "inverse" is used because of the sensor and other real-time measurements we make; it encompasses what we really want to know about what is happening "inside" the system, be it inside a machine or a retail outlet or a city. Clearly, this is a very difficult task. Fortunately, the ML and AI community has started to focus on causality and methods are being developed. We have developed our own algorithm (which will be explained in Part II) with the specific objective of capturing the dynamics of the underlying system that generated the sensor and other real-time surface data.

In the next chapter, we discuss some basic ML tools and applications to a simple problem (the classification of the Fisher Iris flowers). This will familiarize the reader with supervised and unsupervised learning, a few ML algorithms, and MATLAB code to actually do the classification and thus build up first-hand familiarity with ML.

REFERENCES

[CJ13] Casti, J, *X-Events: The Collapse of Everything*, William Morrow, 2013.

[DR73] Duda, R and Hart, P, *Pattern Classification and Scene Analysis*, John Wiley, 1973.

[MP97] Madhavan, PG, *Instantaneous Scale of Fluctuation Using Kalman-TFD and Applications in Machine Tool Monitoring*, SPIE Proceedings, 1997.

INTRODUCTION TO MACHINE LEARNING

This chapter will address the basics of machine learning and explain the steps used in finding a solution to a problem using machine learning.

BASIC MACHINE LEARNING

Figure 2.1 shows a basic taxonomy of machine learning (ML) used by MATLAB for its machine learning toolbox.

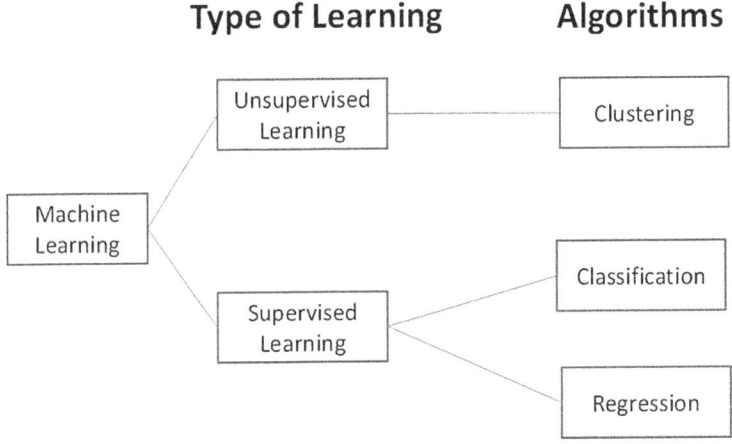

Figure 2.1 Machine learning taxonomy (adapted from Mathworks Inc.)

If we take a more unified approach to this taxonomy, we can achieve more:

- When a general framework is fully described, the various tools in ML can be seen as special cases.

- It will become clear what assumptions about the data and peculiarities of the applications lead to special cases.

- If the assumptions fit, use the special case since, in general, special cases will be more finely tuned for applications where the assumptions fit.

Under supervised learning, classification and regression can be seen as the same at one level with just the output being different.

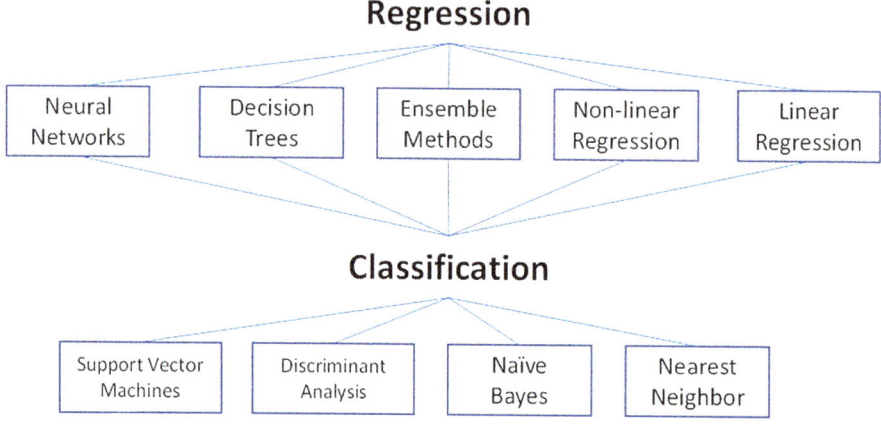

Figure 2.2 Types of regression and classification methods (adapted from Mathworks Inc.)

Classification is regression with outputs mapped into integers. Classification is discrete-valued and regression is continuous-valued. Typically, classification outputs can be names or labels or categorical variables. In the Iris data case, the three classes are "Iris-setosa," "Iris-versicolor," and "Iris-virginica." There is no reason why we cannot use a mapping as shown in the Table 2.1.

Regression outputs can be constrained to integers or we can use a simple rule such that Class 1 = 1.0±0.5, Class 2 = 2.0±0.5, and Class 3 = 3.0±0.5.

Table 2.1 Classification Mapping

Class Name	Mapped Class Name
Iris-setosa	1
Iris-versicolor	2
Iris-virginica	3

Since classification problems can be cast as regression problems, let us focus on regression in the study of supervised learning.

Regression

i. There are many types of regression.
Simple linear regression
Remember the straight line equation:

$$y = mx + c$$

where y is the class and x is the attribute, m is the slope, and c is the y-axis intercept of the linear relationship or map. In the supervised ML context, this means given a large number of (x, y) pairs, learn m and c.

ii. Multiple linear regression

$$y = a_0 + a_1 x_1 + a_2 x_2 + \cdots + a_i x_i + \cdots + a_N x_N$$

Instead of one attribute, there are N attributes to describe the classes. Again, in the supervised ML context, this means to learn the $(N+1)$ coefficients, $[a_0 a_1 \cdots a_N]$.

iii. Some generalization of notation.

- Each x can be a vector. Then this is a multi-input, single-output (MISO) system.

- y can also be a vector. That is a MIMO system.

- If $[x_0 x_1 \cdots x_N]$ are delayed versions of x, and x's are random variables, this is a time series stochastic model (of the moving average type) with random noise, ε, added.

$$Y = a_0 + a_1 x_1 + a_2 x_2 + \cdots + a_i x_i + \cdots + a_N x_N + \varepsilon$$

   - State-space equation: Generalizing the time series model, the state-space equation gives a context sensitive formulation.

Vector, $X(t)^T = [x_0 x_1 \cdots x_N]$ with (n) added to show discrete time-dependence explicitly,

$$S(n+1) = A*S(n) + B*X(n) + \varepsilon(n)$$

$$y(n) = C*S(n) + \nu(n)$$

Here, S is a new model parameter introduced: states. A, B, and C are appropriately dimensioned matrices or vectors.

○ Markov Model: Compared to the state space, the first-order Markov property governs the state-transition equation:

$$S(n+1) = A*S(n) + \varepsilon(n)$$

$$y(n) = X(n)*S(n) + \nu(n)$$

iv. Non-linear Regression

When the equation is linear in the parameters, it is called a linear regression. For example,

$$y = a_0 + a_1 x^2 + a_2 x^{1.5}$$

is a case of linear regression.

However, if the coefficients $[a_0\, a_1 \ldots a_N]$ appear in the equation as a non-linearity, it is a non-linear regression case. For example,

$$y = e^{-x} + \varepsilon$$

is a non-linear regression problem because y is related to x via exponentiation.

All machine learning solutions follow the same steps.

Figure 2.3 The machine learning process

Feature extraction is an optional step.

- In many cases, features = attributes.

- However, if higher-order (such as being more information bearing or less noisy) features can be extracted from the attributes (a data mining step), it will greatly enhance the regression step both during learning and generalization (the actual use of the model).

For our exposition of basic ML techniques, we use the commonly available Fisher IRIS data set (this data set is also included with the companion files to the book or by writing to the publisher at *info@merclearning.com*).

The Iris dataset has four attributes (sepal length, sepal width, petal length, and petal width) and three labelled classes (setosa, versicolor, and virginica).

This dataset is used for supervised learning, so we ignore the class labels and treat this problem as an unsupervised learning example.

NORMALIZATION

Zero-mean; Unit Variance: The following attributes are organized as a matrix.

$$X = \begin{bmatrix} x_{11} & x_{12} & \ldots x_{14} \\ x_{21} & x_{22} & \ldots x_{24} \\ x_{150,1} & x_{150,2} & \ldots x_{150,4} \end{bmatrix}$$

In the Iris dataset case, there are 150 rows of measurements and four attributes for each. X is a (150×4) matrix. For each column, j, find the mean and standard deviation. Then, for each element in column j, subtract the mean and divide by the standard deviation of column j. Do this for all four columns.

From here on, the zero-mean, unit-variance matrix, X, is used for the rest of the processing.

DATA EXPLORATION

Data exploration involves plotting the data in different ways to get some familiarity with the behavior of attributes. It may also include calculating various basic data statistics.

Plotting

Try plotting the data in multiple ways to see what the data tell you. Since there are multiple attributes, let us look at some scatter plots.

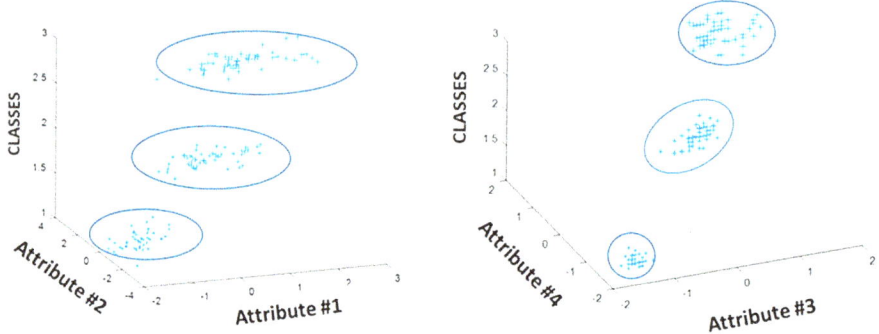

Figure 2.4 Scatter plots of features

The scatter plots of two attributes against the class labels are shown in Figure 2.4, and we can see the classes separated in the vertical direction. However, this does not mean that the classes are easily separable.

In the scatter plot in Figure 2.4 on the left, Attributes #1 and #2 overlap (when you look down from the top), so separating classes in an unknown case using just these 2 attributes can be difficult. Do all the other pairwise scatter plots to get a better sense of how the attributes are divided by classes. The scatter plot for the final pair is shown on the right in Figure 2.4. Class #1 (the class at the bottom) is well-separated. Classes 2 and 3 also seem separated, but may have some small overlap.

PARALLEL COORDINATE SYSTEMS

When there are more than two attributes, the plots in Figure 2.4 indicate that we have to do multiple pairwise plots. There is a way to visualize more than two attributes simultaneously.

Instead of the coordinate axes being orthogonal to each other, why not draw the axes parallel to each other? Then you can draw any number of axes and see the relationships simultaneously.

Each of the three classes are plotted in a different color. For attributes #3 and #4, the three color bands are separated (the same observation from the earlier scatter plot, but perhaps a bit more explicit). As noted before, Classes 2 and 3 seem uncomfortably close. For Attributes #1 and #2, all three classes overlap and there is no simple way to separate the classes.

In 1977, Inselberg advanced the current version of parallel coordinate system. Beyond visualization, some geometry can be developed based on the parallel coordinate representations.

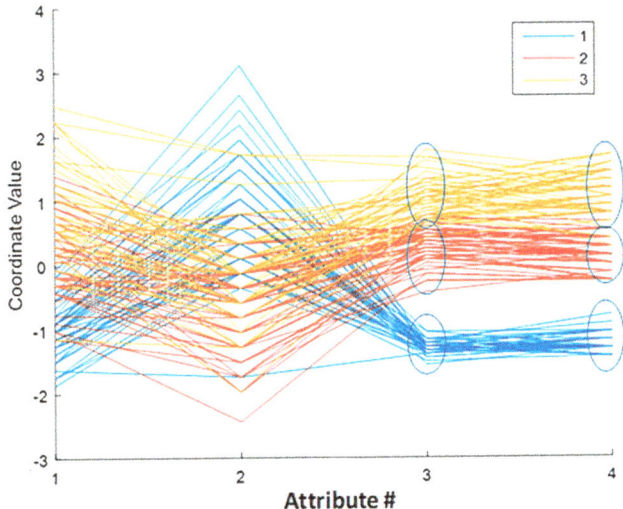

Figure 2.5 Parallel coordinate plot

For example, a straight line on a plane in the Cartesian coordinate system appears as a collection of line segments between two parallel coordinates that all cross at a point (between the two parallel axes or outside or at infinity). Additional geometric developments were found to be hard to do.

FEATURE EXTRACTION

Feature extraction is a desirable step that will improve the following stages of ML. Attributes or raw data may have noise (anything unwanted, in general); also, they may not be represented in the most efficient manner.

One approach is to transform the attributes that are vectors represented on a standard basis into one represented on its own eigenvector basis. This may allow dimensionality reduction, noise subspace removal, and the most compact representation possible for the attributes.

From linear algebra, you know that from the data matrix (of attributes) and its correlation matrix, eigenvalues and vectors can be calculated.

$$X = \begin{bmatrix} x_{11} & x_{12} & \cdots x_{14} \\ x_{21} & x_{22} & \cdots x_{24} \\ x_{150,1} & x_{150,2} & \cdots x_{150,4} \end{bmatrix}$$

The correlation matrix, $R = X^T X$, a (4×4) matrix.

For the Iris data, correlation matrix,

$$R = \begin{bmatrix} 0.9933 & -0.1086 & 0.8659 & 0.8125 \\ -0.1086 & 0.9933 & -0.4177 & -0.3542 \\ 0.8659 & -0.4177 & 0.9933 & 0.9563 \\ 0.8125 & -0.3542 & 0.9563 & 0.9933 \end{bmatrix}$$

This symmetric matrix has as diagonal elements, 0.9933 (which should have been equal to 1, since we had made the data unit variance in the normalization step yet close enough to 1). The off-diagonal entries are non-zero, which indicates that the four attributes are correlated (meaning they present similar information, but this redundancy can be removed).

The results of the eigen-decomposition in MATLAB are shown in Figure 2.6.

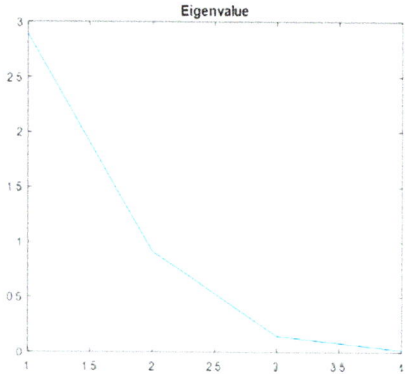

Figure 2.6 Eigenvalue plot results of the eigen-decomposition in MATLAB

In the eigenvalue plot in Figure 2.6, we can find that:

- The X-axis consists of the four eigenvalues.
- The Y-axis shows a value of each one.

- The 4th eigenvalue ≈ 0.

- This is an indication that we can represent the attributes effectively with three instead of four numbers. In other words, three features are sufficient instead of using four attributes in the subsequent steps.

- In fact, for reasons mentioned above, it is better to use the three features because they have nice properties.

This also means that we choose the first three eigenvectors as the basis for representation, and the features so obtained will be a better choice than the original four attributes.

For feature matrix, $F = X*Q$,

- Q is (4×3) a reduced eigenvector matrix that contains only the first three eigenvectors corresponding to the three largest eigenvalues.

- F is a (150×3) feature matrix derived from the attribute matrix, X, which is (150×4).

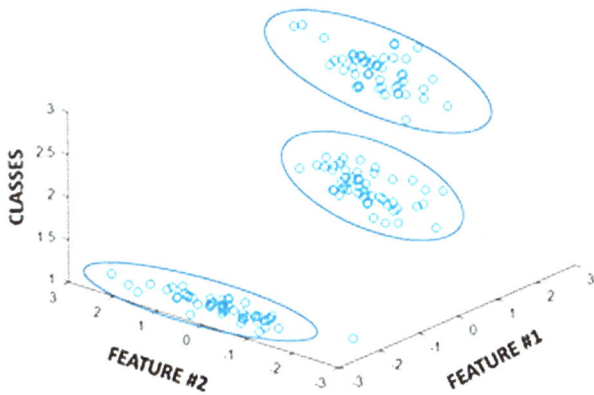

Figure 2.7 Feature plot of three labelled classes

In the scatter plot of Figure 2.7 on the right, two out of the three Features (all 150 of them) are plotted against the three labelled classes. This scatter plot is that of the features and the previous one was of attributes, but it is clear that the class clouds here are "tighter" in the case of the feature vectors. This is the visual effect of eigen-decomposition and the steps we took.

$$\text{Correlation Matrix of Features} = \begin{bmatrix} 2.8914 & 0.0000 & 0.0000 \\ 0.0000 & 0.9151 & 0.0000 \\ 0.0000 & 0.0000 & 0.1464 \end{bmatrix}$$

The feature correlation matrix is diagonal, proving that features are not correlated (unlike the attributes) and this property is very beneficial in the subsequent ML steps.

However, when we use these three features, we lose the ability to make practical interpretations. Remember that the four attributes, the sepal length, sepal width, petal length, and petal width, had physical meanings. When converted into the three features, the features have no physical meaning (in most cases). So, by using the features, our math will be better (for example, there will be no uncorrelated features) and we can expect better regression results. However, we should be willing to pay the cost of lost interpretability for the advantage of less errors.

Regression

We can perform the following regression methods using the Iris data:

1. Linear Regression

2. Decision Tree

3. Naïve Bayes

In this example, the Iris data are divided into

a. Training set – random 90 data samples

b. Test set – random 60 data samples

The three regression methods are trained using the Training Set (class labels available) and tested on the Test Set (class labels not available). The initial diagnostics/ performance is assessed using the

a. Training Error (sometimes called the *prediction error*) and

b. Test Error (sometimes called the *generalization error*).

Then we combine the three regression methods to create an ML solution and perform various diagnostics.

MULTIPLE LINEAR REGRESSION

For this example, use MATLAB program *fitlm* and use the simple linear least squares algorithm. The regression equation that is generated is

Class = 2 + 1.9953*Feature1 − 0.026357*Feature2 − 0.4436*Feature3

Training Error, MSE = 4.86%

Now, you can use this regression equation and use Test Set Features 1, 2, and 3 to obtain the class labels that this regression equation predicts.

The class labels generated are plotted for the 60 Test Set Features. Red indicates the actual outputs of the regression equations and the Test Error MSE = 4.45%.

It may seem odd that Test Error is less than Training Error; after repeated random tests, ensemble averages of the two MSEs will show the expected behavior, i.e., Test MSE > Training MSE.

A simple variation in the outputs generated by the regression equation provides class labels that are more interpretable. Remember that the class labels are "1," "2," and "3." Using a simple banding, the regression equation outputs were constrained to be integers. They are plotted in blue in Figure 2.8. One class "2" and two Class "3" test samples where misclassified, whereas all Class "1" samples where correctly classified.

The Test Error for the modified outputs is MSE = 5%.

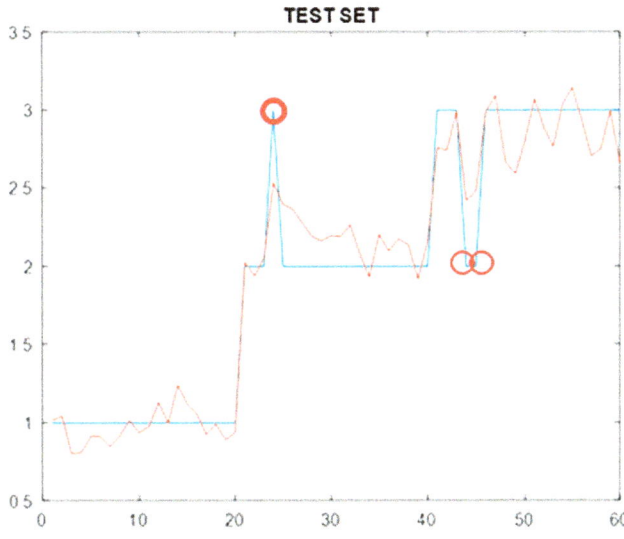

Figure 2.8 Multiple linear regression results

This concludes the multiple linear regression model exploration.

DECISION TREE

For this example, use MATLAB program *fitrtree* and use the FITRTREE algorithm, which grows the tree using the MSE (mean squared error) as the splitting criterion. At each node of the tree, choose the attribute that most effectively splits its set of samples into subsets enriched in one class or the other.

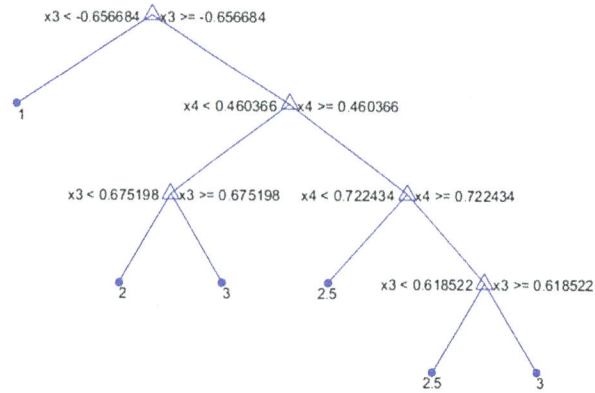

Figure 2.9 Converged decision tree for the Iris training data

The Training Set is shown in the converged tree in Figure 2.9.

In this example for the Iris dataset,

- Attribute $x1$ = Sepal length
- Attribute $x2$ = Sepal width
- Attribute $x3$ = Petal length
- Attribute $x4$ = Petal width

Leaves are the classes:

- Class 1 = setosa (=1)
- Class 2 = versicolor (=2)
- Class 3 = virginica (=3)

Training Error, MSE = 1.67%

When the Test Set was applied to this tree, the Test Error MSE = 1.25%.

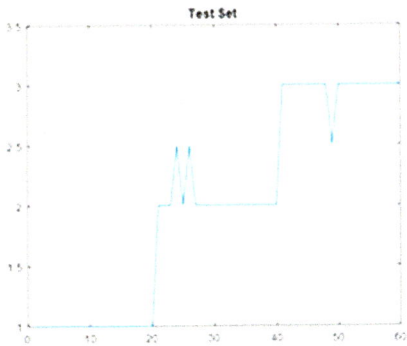

Figure 2.10 Decision tree results for the Iris test data

Since regression gives numerical values which we had not constrained to be integers, Figure 2.20 shows that the three test samples were classified as Class "2.5." A simple resolution is to allocate each "2.5" output to a Class "2" or Class "3" arbitrarily.

The major attraction of decision trees is that we can state in plain English why unseen samples were classified as setosa, versicolor, or virginica. The rules can be written out from the tree diagram (Figure 2.9) by inspection (shown in Figure 2.11).

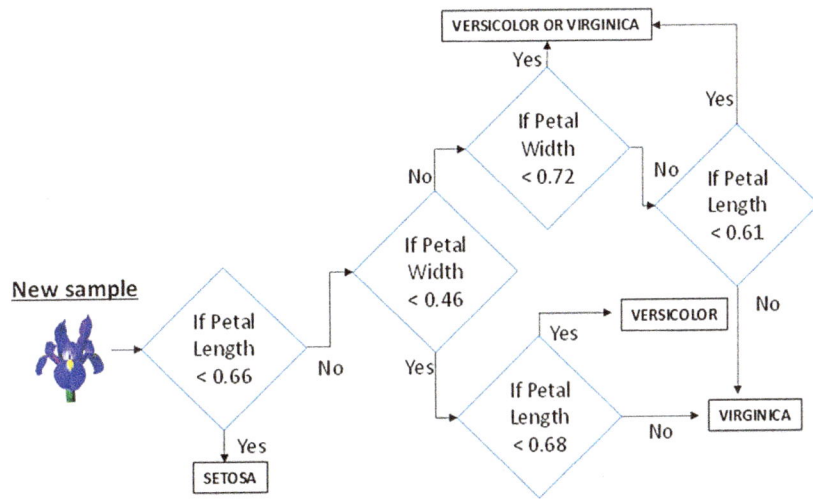

Figure 2.11 Flowchart from the decision tree given in Figure 2.9

We can easily see why a decision to call a flower "setosa" was reached. We also notice that the sepal length and width do not play a part in the decision making. Note that the attribute inputs are normalized values.

NAÏVE BAYES

For this example, use the MATLAB program *iris_NB* and the open-source code included the book's companion files. Use the Simplified Naïve Bayes algorithm.

Since the "naïve" portion of Naïve Bayes expects input features to be uncorrelated, the three features from the Multiple Linear Regression section instead of the four attributes are used.

We model each feature with a Gaussian (or kernel or multinomial) distribution. Then, the Naïve Bayes classifier estimates a separate normal distribution for each of the three classes by computing the mean and standard deviation of the feature training data in that class.

The Test Set results are shown in Figure 2.12. Clearly, many test samples are misclassified. Indeed, for the Test Error, the MSE = 48.33%.

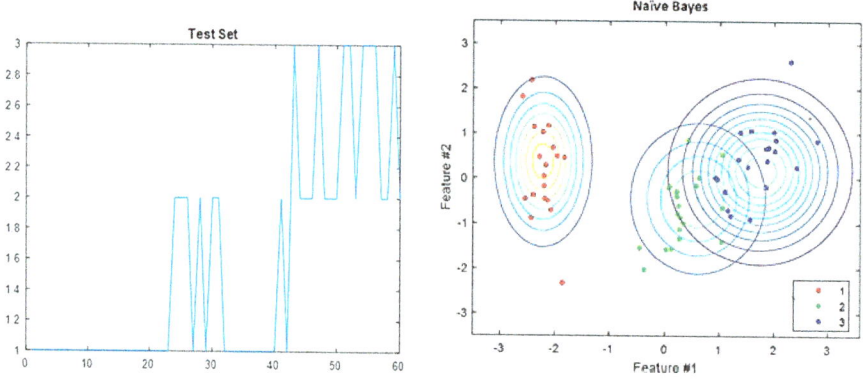

Figure 2.12 The Naïve Bayes Test Set results

Figure 2.12's results are not indicative of the Naïve Bayes performance; it is simply that we have not optimized the code for the Naïve Bayes implementation. The three Gaussian distributions with different centers and standard deviation axes were learned using the Training Set (shown on the right in Figure 2.12). The Test Set attributes (3 and 4) are plotted as dots on the same figure. The three Gaussians learned earlier capture the Test Set attributes with multiple overlaps.

ENSEMBLE METHOD

For this example, use the MATLAB program *iris_SB* and the open-source code in the book's companion files, as well as the Simple Boosting algorithm. Previously, you saw three methods. How do we combine them so that we obtain better results?

1. **Boosting:** This generates some "weak" learners (say, simple linear regression) and adaptively combines them to classify complicated class separators (such as a circular class inside another annular class).

2. **Bagging:** This uses the Bootstrap method (resampling with replacement), generates multiple results, and takes their average.

In this section, we utilize our own Ensemble Method using Multiple Linear Regression on already trained classifiers. Note that the following is not the Adaboost algorithm developed by Freund and Schapire (1995).

Remember from our simple Multiple Linear regression that the following is true:

$$y = a_0 + a_1 x_1 + a_2 x_2 + \cdots + a_i x_i + \cdots + a_N x_N$$

where all values of $\{x\}$ are the attributes and all values of $\{y\}$ are the predicted class labels.

Let us assume we have M such regression models (or other ML models) that provide M different predicted class labels. Now, we can create a prediction of the class labels, Y, using $\{y_1, y_2, \ldots, y_M\}$.

$$Y = b_0 + b_1 y_1 + b_2 y_2 + \cdots + b_i y_M$$

We solve this problem: (find $\{b\}$) using the Least Squares (pseudo-inverse) method. In Adaboost, a "weighted recursive least squares" procedure is used for the solution, so it is called "Ada" (short for "Adaptive").

Combining the M regressors via another multiple linear regression (or classifiers) should make "Y" a better prediction of the class labels than any individual $\{y\}$.

We combine our previous Multiple Linear Regression (MLR) and Decision Tree (DT) models (actually, outputs) using our Simple Boost (SB) method.

$$Y_{SB} = b_0 + b_1 y_{MLR} + b_2 y_{DT}$$

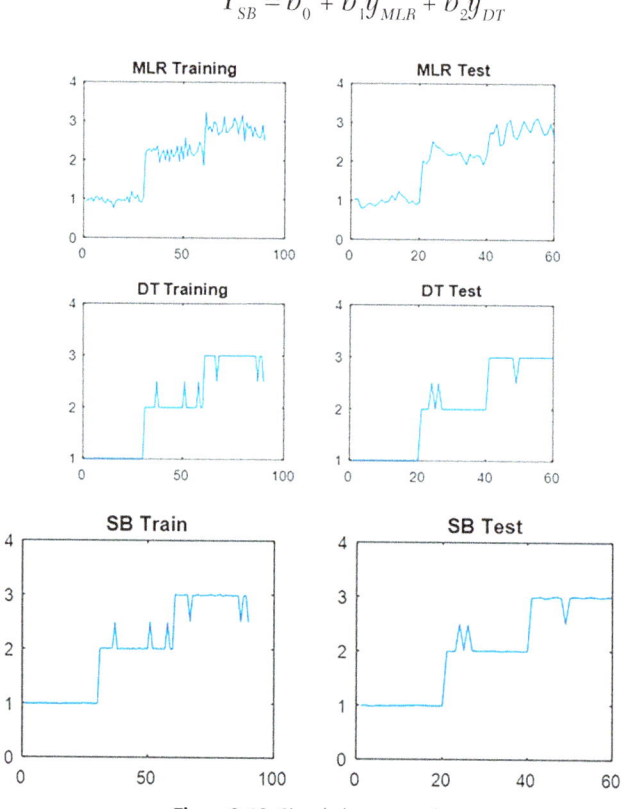

Figure 2.13 Simple boost results

The results of training and testing are shown in Figure 2.13. The SB regression coefficients are

$$Y_{SB} = -0.0048 + 0.0463 \, y_{MLR} + 0.9561 \, y_{DT}$$

From the coefficients, it is clear that the SB model is heavily weighted in favor of the DT model (the coefficient is high). So, we expect SB model predictions to be close to the DT model predictions, which is shown in the figure. The Training Error MSE = 1.66% and the Test Error MSE = 1.24%, which are both very close to the DT MSEs.

In an adaptive boost solution, for points 37 or 51 (and four more) in the DT Training set, the weighting given to SB would have been much lower so that for the SB, so the prediction would have been closer to that of the MLR.

UNSUPERVISED LEARNING

Unsupervised learning is challenging because there are no Class Labels available for training.

Unsupervised Learning

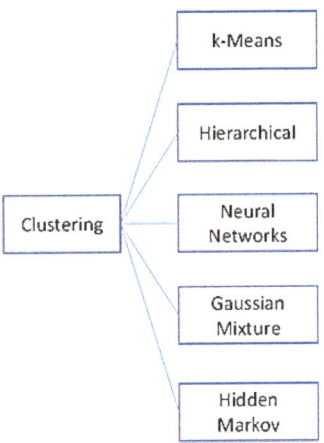

Figure 2.14 Unsupervised learning methods (adapted from Mathworks Inc.)

However, there are certain things we can do to make progress. In general, the approach is to group the attributes (or features) in some fashion so that some "global" desirable property improves. An example of a global desirable property could be that the groupings obtained have a minimum correlation between each other or that the overall information content is minimized (so that there is minimum redundancy). These global properties can be quantified and an algorithm created to get the best groupings. Five such methods are shown in Figure 2.14. We consider two of them.

To reuse the Iris dataset, assume that the class label column is not available in the attribute dataset. Once the clusters are obtained, we compare them with the class labels we had held back to get accurate measurements. In other words, the attribute dataset is treated as the Test Set.

K-MEANS CLUSTERING

In this example, we use the MATLAB program *kmeans* and the k-Means algorithm.

Given k centers, the k-Means algorithm minimizes the within-cluster sum of the squares of the distance functions of each point in the cluster to the k centers. There are many distance functions, and the simplest among them is the Euclidian Distance. The k-Means algorithm does the minimization iteratively (and can get stuck in local minima).

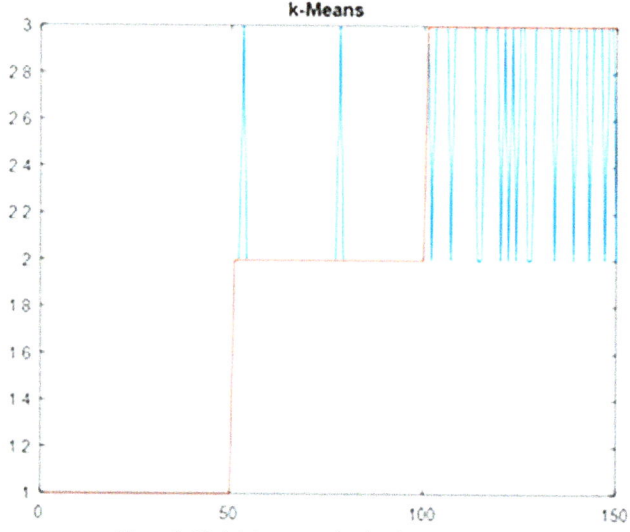

Figure 2.15 k-Means results for the Iris dataset

Using all 150 attributes (for this run, we used un-normalized attributes), the k-means algorithm needs to find the number of clusters expected. In this trial, we gave "3" as an input. Our feature extraction step gave us that clue. The results are shown in Figure 2.15 (clustering error MSE = 10.67%).

 For each run of the k-means algorithm, the starting points are different, and hence you may see highly variable results based on which local minima the algorithm got stuck in.

SELF-ORGANIZING MAP (SOM) CLUSTERING

For this example, we use the MATLAB program *somtoolbox* and the open-source code at *www.cis.hut.fi/projects/somtoolbox/*. We also use the Kohonen algorithm.

SOM is a neural network-inspired clustering solution, quite unlike back-propagation artificial neural networks. SOM is a group of "neurons" organized on a low-dimensional grid. Each neuron is represented by a weight vector, m, with dimension $(px1)$ where p = number of attributes (which equals 4, in the Iris examples). Neurons are connected to neighboring neurons in a hexagonal lattice structure with neighborhood relationships. As training proceeds, neurons on the grid become ordered with neighboring neurons having similar weight vectors.

Each attribute vector is considered one at a time. Each neuron's vector, m, is compared to the current attribute vector and the closest neuron is called the Best Matching Unit (BMU). Closeness can be measured via Euclidean or other distances. Topological neighbors of BMU are updated so that they come closer to BMU (such that their m vector is closer to that of BMU's).

For each attribute comparison step, n, for neuron j, update its weight vector by the following:

$$m_j(n+1) = m_j(n) + \alpha(n)\, h_{cj}(n)\, [x(n) - m_j(n)]$$

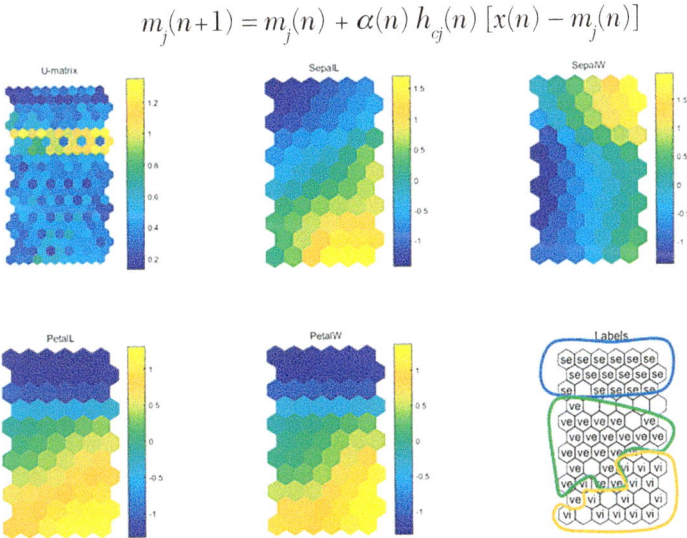

Figure 2.16 Converged SOM results

where x is the attribute vector, h is the neighborhood kernel around BMU, c, and α is the learning rate.

The clustering accuracy is 1.3%, which is remarkable for unsupervised learning. As you see in the bottom-right in Figure 2.16, the three types of irises are nicely segregated into different regions.

Now that we have identified multiple supervised and unsupervised methods of machine learning, the last step is to apply them to a suitable business problem and create some actionable recommendations. Consider the following example.

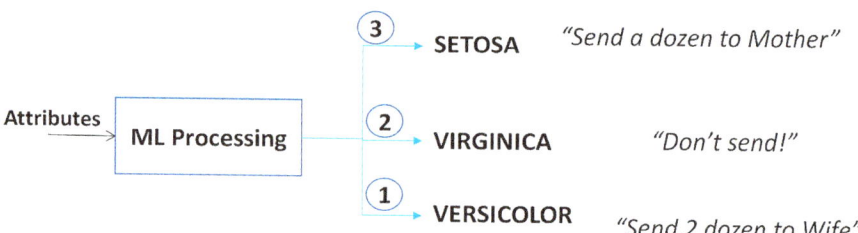

Figure 2.17 A business use case for clustering

CONCLUSION

This chapter was a brief introduction to machine learning. You gained familiarity with multiple ML techniques under an organized framework. Now you are ready to attend a good online course on ML, understand the theory behind these models, write your own code in MATLAB or R or Python, and work with large datasets.

CHAPTER 3

SYSTEMS THEORY, LINEAR ALGEBRA, AND ANALYTICS BASICS

S ignals and systems, linear algebra, and digital signal processing (DSP) are typically encountered in the undergraduate years and have much in common with the ML basics introduced in the last chapter. This chapter looks at their commonality and presents them using the framework of linear algebra. At times, rigor is sacrificed in favor of insights, motivation and unifying viewpoints.

DIGITAL SIGNAL PROCESSING (DSP) MACHINE LEARNING (ML)

The basic building block of DSP is a filter, which is depicted in Figure 3.1 as a tapped delay line.

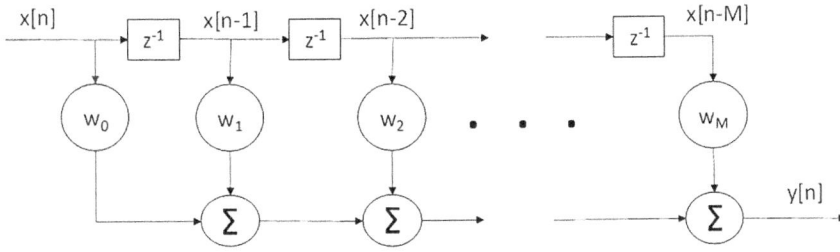

Figure 3.1 Tapped delay line

In Figure 3.1, $x[n]$ is the input at discrete time index, n; z^{-1} is the delay operator such that its output is a one-unit delayed version of its input, the values for w are the weights, and $y[n]$ is the output.

$$y[n] = \sum_{k=0}^{M} w_k \, x[n-k]$$

By appropriately choosing values for w, $y[n]$ becomes a filtered version of $x[n]$ with desired properties. The values for w can be chosen a priori if the properties of $x[n]$ and $y[n]$ are known; if not, they can be generated adaptively. *It is this latter "learning" approach that is the basis of the equivalence between DSP and ML.* We notice that the equation above is the same as that of the Multiple Linear Regression equation in the ML basics chapter. Now, the tapped delay line or the summation equation can be written as the appropriate matrix and vectors.

If we consider a **MIMO** (multiple input, multiple output) version with $(M + 1)$ inputs and $(P + 1)$ outputs ($P = 0$ yields the tapped delay line), we obtain

$$y - a \ (P + 1)x1 \text{ vector}; \ x - a \ (M + 1)x1 \text{ vector},$$

which yields a matrix that has the w variables arranged in a special way, a $(P + 1) \times (M + 1)$ matrix.

$$y = Ax$$

$$\begin{bmatrix} y[n] \\ y[n-1] \\ y[n-2] \\ \dots \\ \dots \\ y[n-P] \end{bmatrix} = \begin{bmatrix} W_0 & W_1 & \cdots & \cdots & \cdots & W_{M-1} & W_M \\ 0 & W_0 & W_1 & & & & \\ 0 & 0 & W_0 & & & & \\ & & & & & & \\ 0 & & & 0 & & & W_{M-P} \end{bmatrix} \begin{bmatrix} x[n] \\ x[n-1] \\ \dots \\ \dots \\ \dots \\ y[n-M] \end{bmatrix}$$

This is a key vector-matrix equation that we will see again and again as we transition to the linear algebraic formulation.

LINEAR TIME INVARIANT (LTI) SYSTEM

The tapped delay line filter we saw in the previous section is a Linear Time Invariant system.

A VARIETY OF OBSERVATIONS CAN BE MADE:

- In DSP, the foregoing topics have been studied for many decades with insights from the use of Transform theory (Fourier and Wavelets, for example), filter design, matrix factorization, polynomial theory, and adaptive filters.

- In systems theory, the pattern of entries of "A" matrix is revealing (linear time-invariant systems). The associated rich heritage of system control, stability, and optimality brings more power into ML.

- It will become apparent that such systems constitute a commutative algebra. Pure mathematicians can see connections to group theory and realization theory.

A major objective of our exercise here is to accelerate the deepening of ML theory and adoption of existing powerful solutions in related domains.

Considering a discrete-time $\underline{x}$ and $\underline{y}$ similar to the previous section, but dissimilar to tapped delay line summation equation (the YET equivalent in the Transform domain), the following algebraic equation relates $\underline{y}$ to $\underline{x}$. Taking Z-transforms (z being a complex variable of the form, $z = \overline{c} + j*d$, where $c, d \in R$).

$Y(z) = A(z) \, X(z)$, where each term is the Z-transform of its time counterpart.

$A(z)$ is of the form of a ratio of polynomials,

$$A(z) = \frac{q_0 + q_1 z + q_2 z^2 + \cdots + q_M z^M}{1 + p_1 z + p_2 z^2 + p_3 z^3 + \cdots + p_N z^N}$$

$$= h_0 + h_1 z + h_2 z^2 + \cdots + h_R z^R$$

where $\{h\}$ is called the *impulse response*.

The roots of the numerator and denominator polynomials have great meaning in LTI system analysis. The roots of the numerator are zeros, and the roots of the denominator are poles. For the practitioner, the locations of the poles and zeros in the z-plane describe the system's performance. For example, if all the zeros are outside the unit circle in the z-plane, the system is invertible and is called a *minimum-phase* system.

When $A(z)$ is evaluated on the unit circle in the z-plane, $A(z)$ is the Fourier transform of A, written as $A(k)$, where k stands for the frequency sample number. On equally spaced samples on the unit circle, $z = e^{-j\frac{2\pi}{N}k}$, where N is the total number of samples. The discrete Fourier transform is

$A(k) = \sum_{n=0}^{N-1} h[n] e^{-j\frac{2\pi}{N}kn}$. The Fourier transform of A, $A(k)$ in the summation equation above can also be written in matrix-vector form:

$$\underline{A} = \underline{E}\,\underline{h}$$

$$\begin{bmatrix} A(0) \\ A(1) \\ \cdots \\ \cdots \\ A(N-1) \end{bmatrix} = \begin{bmatrix} 1 & 1 & \cdots & 1 \\ 1 & \exp(-j2\pi/N) & & \exp(-j2\pi(N-1)/N) \\ 1 & \exp(-j2\pi(N-1)/N) & & \exp(-j2\pi(N-1)^2/N) \end{bmatrix} \begin{bmatrix} h(0) \\ h(1) \\ \cdots \\ \cdots \\ h(N-1) \end{bmatrix}$$

The complex exponential vector matrix, $\underline{E}$, is a special matrix. Its column vectors are eigenvectors.

For LTI systems, complex exponentials are eigenfunctions. This means that if the input is a complex exponential, the output will also be a complex exponential (with scaling) if A is LTI.

One last observation in this section is regarding a special type of random input and time series analysis. If the elements of $\underline{x}$ are random variables which are independent and identically distributed (i.i.d.), the output is called a *random time series of the ARMA type*. The auto regressive (AR) time series model has the numerator polynomial of $A(z) = 1$ and **MA** or moving average time series model has the denominator polynomial of $A(z) = 1$. The combination is an **ARMA** model and variations of ARMA are also defined (such as **ARIMA**, where "I" stands for "Integrated," and it is used for non-stationary time series analysis).

In the previous two sections, $\underline{x}$ had a particular structure. Elements of $\underline{x}$ were delayed versions of $x[n]$ or Time Samples or Observations ordered over Time. Now, time happens to be an example of an independent variable

---

**WELL-KNOWN FACTS
RECOUNTED IN THIS SECTION**

- By pulling together the Z transform, poles and zeros, the Fourier transform, and eigenfunctions/ vectors, we can understand how they are related.

- The magnitude of the Fourier transform of $\underline{A}$ is the filter-band characteristics (low-pass) and the magnitude squared is an estimate of the power spectral density of the ARMA time series; all of it can be visualized if we know the pole and zero positions in the z-plane.

for ordering. All the previous results hold if observations are ordered over any other independent variable.

Consider the surface of a lake and the height of the waves at all points as our observations. They are ordered over the values (x, y), creating a two-dimensional plane. Another example is an image whose grey-level varies over two dimensions. Everything we discussed so far is still true. In the case when we are dealing with space, a few notions have to be kept in mind. Frequency is the number of cycles over the unit length. How the frequency is perceived is different depending on the direction you are looking at. For example, assume you are standing on the shore and the waves are rolling in evenly in a parallel fashion, and you are looking out straight ahead to the horizon. In that direction, you can see the distance between the wave crests at any instant and that gives you a certain spatial frequency of (for example) three waves per meter. If you turn 90° and look along the shore, you see just one crest all along the shore of a long wave – the frequency is 0 cycles/meter. As you turn your eyes from along the shore to the horizon, the spatial frequency goes from 0 to 3 cycles/meter. This is called frequency of corrugation. When we go from one to two dimensions, the concept of frequency becomes more elaborate. Now, the observations do not need to be ordered over time or space. We can use any meaningful ordering principle. This notion in often useful in ML. Let's consider the Iris classification exercise in the last chapter. The attributes for each observation were the sepal length, sepal width, petal length, and petal width. If we maintain the same ordering, we can apply the tools from this chapter. For example, a new feature set derived from these attributes for the Multiple Linear Regression example could be developed as follows.

$$\begin{bmatrix} A(0) \\ A(1) \\ \cdots \\ \cdots \\ A(N-1) \end{bmatrix} = \begin{bmatrix} 1 & 1 & \cdots & 1 \\ 1 & \exp(-j2\pi/N) & & \exp(-j2\pi(N-1)/N) \\ & & & \\ 1 & \exp(-j2\pi(N-1)/N) & & \exp(-j2\pi(N-1)^2/N) \end{bmatrix} \begin{bmatrix} h(0) \\ h(1) \\ \cdots \\ \cdots \\ h(N-1) \end{bmatrix}$$

$N = 4$; $\{h(0), h(1), h(2), h(3)\}$ = (sepal length, sepal width, petal length, and petal width).

Then $\{A(0), A(1), A(2), A(3)\}$ are the new Fourier features for regression. Try these new features and compare your results to those from the previous chapter.

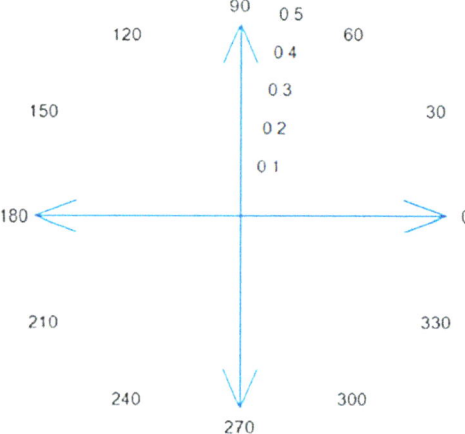

Figure 3.2 Complex exponentials of the E matrix

Here are some points to note about Figure 3.3:

- It shows the polar plot of the complex exponentials of E.

- It shows all 150 "Fourier" features on the right.

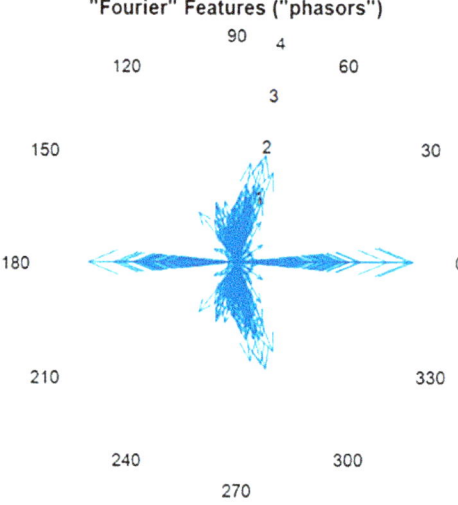

Figure 3.3 Features of the Iris data

- The new features are complex numbers. You will have to be careful when you set up the multiple linear regression solution for the Iris classification.

- You can consider the magnitude or phase of our Fourier features as features to use for classification.

- The results shown in Figure 3.3 only used the phase of the Fourier features.

Instead of using an explicit training and test approach, the least squares solution using pseudo-inverse was used on all 150 data samples to perform this classification exercise since the focus is on using the $A = Eh$ approach to develop a set of Fourier features when the data are not ordered over time. In Figure 3.4.

- Using the feature phases only, the regression outputs are as shown on the left in blue. Compared to the true classes (in red), the classes are well-identified even though the regression output values are not 1, 2, and 3.

- We used the same "banding" approach as shown in the previous chapter to obtain the results on the right.

- Only two irises (around feature 100) were misclassified. The training error was 1.3%.

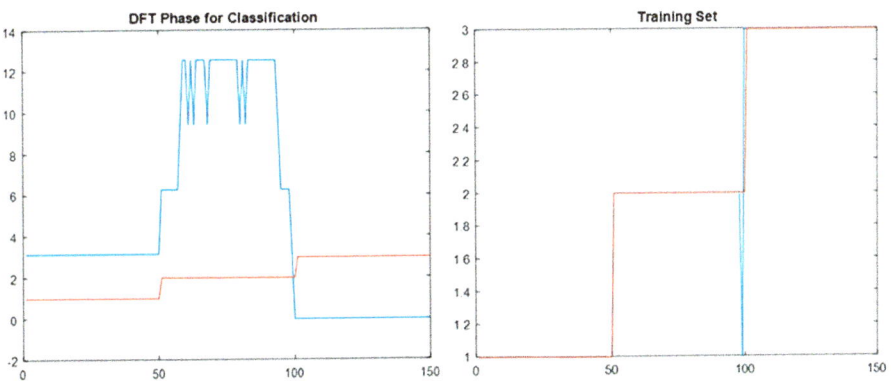

Figure 3.4 Classification results using Fourier transform phase

---

FOURIER TRANSFORM INSIGHTS

- The Fourier transform is complex with real and imaginary parts.

- The magnitude tells you "what" and the phase tells you "how."

- The magnitude tells you what type of sinusoids are present and the phase tells you how they are put together.

- The phase information tends to be ignored, but it can include significant information in applications such as speech and music analysis.

LINEAR ALGEBRA

Linear algebra mechanics are undergraduate topics; in advanced courses, we are introduces students to advanced concepts such as eigenvalues and vectors, various matrix factorizations, basis and projections, and subspace decomposition. These concepts are integrally related to systems and Digital Signal Processing (or DSP). We start with basic linear algebra so that a shared terminology is developed and concepts from other fields can be connected as we go along.

The basic Linear Time Invariant system we considered earlier in this chapter is also a good starting point for linear algebra.

$$\underline{x} \longrightarrow \boxed{\underline{A}} \longrightarrow \underline{y}$$

$\underline{A}$ is an $(M \times N)$ matrix, $\underline{x}$ is a $(N \times 1)$ vector, and $\underline{y}$ is a $(M \times 1)$ vector. In LTI system terms, this represents a MIMO system with N inputs and M output. The output $\underline{y} = \underline{A}\,\underline{x}$.

In linear algebraic terms, this linear transformation can be written as

$$T\colon \underline{x} \rightarrow \underline{y} \text{ where } \underline{x} \in R^N,\ \underline{y} \in R^M$$

This can be read as, "Transformation from $\underline{x}$ to $\underline{y}$ where $\underline{x}$ is a member of the reals of N-dimensional space and $\underline{y}$ is a member of the reals of M-dimensional space (instead of reals, we can also consider complex fields)." This relationship can be written in equivalent forms:

$$T(\underline{x}) = \underline{A}\,\underline{x} \text{ or } \underline{y} = \underline{A}\,\underline{x}$$

To recount terminology from basic linear algebra, consider the matrix $\underline{A}$.

$$\underline{A} = \begin{bmatrix} a_{11} & a_{12} & \cdots & a_{1N} \\ a_{21} & a_{22} & & a_{2N} \\ \cdots & & & \cdots \\ \cdots & & & \cdots \\ a_{M1} & a_{M2} & \cdots & a_{MN} \end{bmatrix} = \begin{bmatrix} \uparrow & \uparrow & \uparrow & \uparrow \\ \underline{a}_1 & \underline{a}_2 & \cdots & \underline{a}_N \\ \downarrow & \downarrow & \downarrow & \downarrow \end{bmatrix}$$

Here, we can see that the following are true:

- $\underline{A}$ is a matrix of N column vectors.

- $C(\underline{A})$ is the column space of $\underline{A}$, the M-dimensional subspace spanned by the N column vectors of size $(M \times 1)$ of $\underline{A}$.

- $R(\underline{A})$ is the row space of $\underline{A}$, the N-dimensional subspace spanned by the M row vectors (not shown above) of size $(N \times 1)$ of $\underline{A}$.

- $N(\underline{A})$ is the NULL space of $\underline{A}$, the N-dimensional subspace spanned by the input vectors, $\underline{x}$, such that $\underline{A}\,\underline{x} = \underline{0}$. In other words, $\underline{A}$ transforms some of the input vectors into the zero-vector.

To obtain a systems insight into this example, consider Figure 3.5.

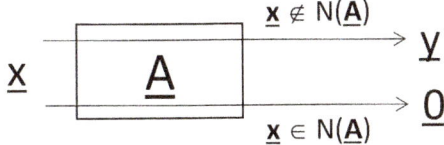

Figure 3.5 A systems view of the example

If $\underline{A}$ represents a low-pass filter, consider two cases:

1. Input, x, has a frequency content within the low pass-band of the filter; then you get some output, y.

2. Input frequencies are high above the filter pass-band; the filter output is zero.

In the first case, $\underline{x}$ lies in $C(\underline{A})$ and in the second, $\underline{x}$ lies in $\mathbf{N(\underline{A})}$.

Now, you can understand matrix properties above in the context of a digital filter.

Let us briefly note that we encountered a special type of input to an LTI system that passes through the system, which was unchanged except for a scaling of the eigenfunctions of the system. There are similar $\underline{x}$ vectors for $\underline{A}$ that are transformed by $\underline{A}$ and emerge unchanged except for scaling. They are the eigenvectors of $\underline{A}$. We will have more to say about them later in the context of the coordinate basis.

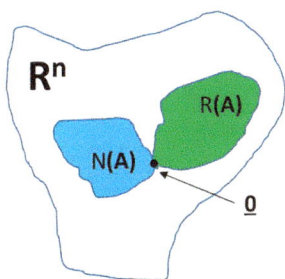

Figure 3.6 The null space

From basic linear algebra, we know the following results (Figure 3.6):

- $N(\underline{A}) \perp R(\underline{A})$ and $C(\underline{A}) \perp N(\underline{A}^T)$, which is called the Left Null Space of $\underline{A}$.
- $N(\underline{A}^T)$ = Left $N(A)$. Also, $R(\underline{A}) \perp C(\underline{A}^T)$.

In general, matrix, $\underline{A}$, transforms vectors from its N-dimensional row-space into its M-dimensional column-space, except for vectors in its null space, which are transformed into the zero-vector.

A Numerical Example

Let us look at a simple example and visualize the various spaces.

$$\underline{A}\,\underline{x} = \underline{y} \text{ with } \underline{A} = \begin{bmatrix} 3 & -2 \\ 6 & -4 \end{bmatrix} \text{ and } \underline{y} = \begin{bmatrix} 9 \\ 18 \end{bmatrix}. \text{ Find } \underline{x}.$$

To directly solve $\underline{x} = \underline{A}^{-1}\,\underline{y}$, $\underline{A}$ has to be invertible. In this case, $\underline{A}$ is not invertible since the columns are linearly dependent (first column $* -2/3 =$ second column).

Using the row reduced echelon form (rref), you get the solution set:

$$\underline{x} = \begin{bmatrix} 3 \\ 0 \end{bmatrix} + c \begin{bmatrix} 2 \\ 3 \end{bmatrix}, \text{ where } c \in R$$

$N(\underline{A})$ is the solution of the homogeneous equation,

$$\underline{A}\,\underline{x} = \underline{0}; \text{ i.e., } N(\underline{A}) = \text{Span} \left(\begin{bmatrix} 2 \\ 3 \end{bmatrix} \right) \quad \underline{A}^T = \begin{bmatrix} 3 & 6 \\ -2 & -4 \end{bmatrix}$$

$$\therefore \text{ Row Space, } C(\underline{A}^T) = \text{Span} \left(\begin{bmatrix} 3 \\ -2 \end{bmatrix}, \begin{bmatrix} 6 \\ -4 \end{bmatrix} \right) = \text{Span} \left(\begin{bmatrix} 3 \\ -2 \end{bmatrix} \right), \text{ since the 2}^{nd}$$

vector is a scaled version of the first.

The solution set is the set of vectors, $\underline{x}$, shown in red. $\underline{x}^*$ in $R(A)$ is the shortest length $\underline{x}$ that can be solution of choice. From the graph, $x^* = \begin{bmatrix} 2 \\ -1.5 \end{bmatrix}$. Additionally, $\underline{x}$ can be seen as a sum of the projections onto the null and row subspaces of $\underline{A}$.

$N(\underline{A})$ is the span of the blue vector and $R(\underline{A})$ is the span of the yellow vector. We also know, $N(\underline{A}) \perp R(\underline{A})$. If $V = N(\underline{A})$ is a subspace of R^2, V^1,

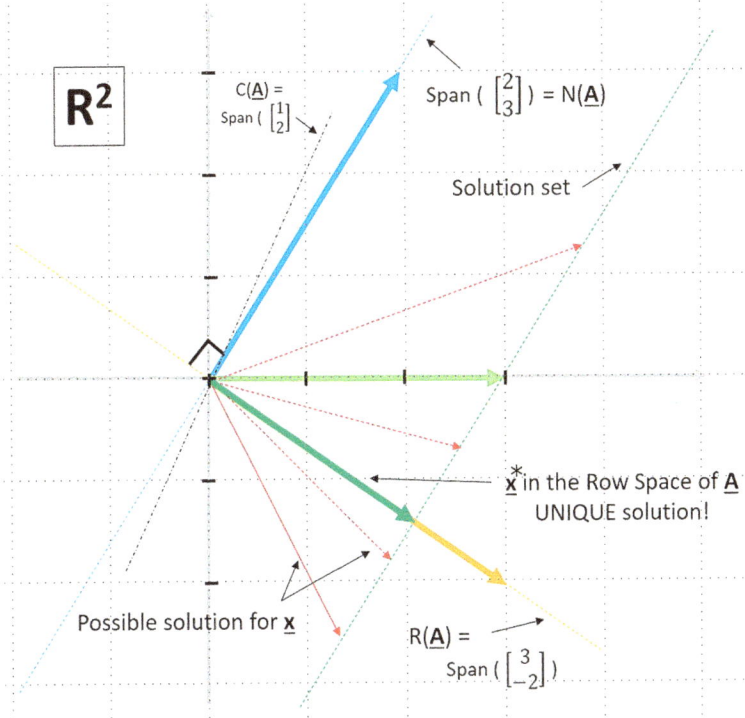

Figure 3.7 Illustration of the column and row spaces; this figure ties together most of concepts discussed earlier in this chapter.

its orthogonal complement is $R(\underline{A})$, another subspace of R^2. If $\underline{s}$ is any of the possible solution vectors (in red) in the solution set, $\underline{s} = \text{Proj}_{N(A)}[\underline{s}] + \text{Proj}_{R(A)}[\underline{s}]$.

The best (shortest) solution, $\underline{x}^*$, is the one where the first term is zero; i.e., $\underline{x} \in$ Row Space of $\underline{A}$. If A is not a square matrix, we have an over- or under-determined system of equations with no unique solution. Then, the least squares solution for $\underline{x}$, $\underline{x}^* = (\underline{A}^T \underline{A})^{-1} \underline{A}^T \underline{y}$.

Projection onto a Subspace

We saw in the last section the projection onto a subspace, where the subspace was a line in R^2. We also saw that we can write $\underline{x} = \text{Proj } V[\underline{x}] + \text{Proj } V_{\perp}[\underline{x}]$, where V is a subspace of R^n and $V^{\perp}$ is its orthogonal complement.

Given V, a subspace of R^n and $\{b_1, b_2, \ldots, b_K\}$, the K basis vectors for V, we can construct a matrix, $\underline{A}$, such that its column vectors are $\{b_i\}$.

$$\underline{A} = \begin{bmatrix} \uparrow & \uparrow & \uparrow & \uparrow \\ \underline{b}_1 & \underline{b}_2 & \cdots & \underline{b}_N \\ \downarrow & \downarrow & \downarrow & \downarrow \end{bmatrix}$$

It can be shown that the projection of $\underline{x}$ onto the subspace, V, $\mathrm{Proj}_V[\underline{x}] = \underline{A}(\underline{A}^T\underline{A})^{-1}\underline{A}^T\underline{x}$. The projection matrix, $\underline{P} = \underline{A}(\underline{A}^T\underline{A})^{-1}\underline{A}^T$. The projection matrix has the property that the values are idempotent, i.e., $P*P = P$.

A Numerical Example

Given a subspace of R^3, $V = \mathrm{Span}\ ([-1\ 1\ 0]^T, [-1\ 0\ 1]^T)$, $\underline{A} = \begin{bmatrix} -1 & -1 \\ 1 & 0 \\ 0 & 1 \end{bmatrix}$.

Then we perform the calculation to obtain the following:

$$\text{Projection matrix, } \underline{P} = \begin{bmatrix} 2/3 & -1/3 & -1/3 \\ -1/3 & 2/3 & -1/3 \\ -1/3 & -1/3 & 2/3 \end{bmatrix}.$$

$\underline{P}$ can be used to project any vector $\underline{x} \in R^3$ to V.

If $\underline{x} = [1\ 2\ 3]^T$; then, $\mathrm{Proj}\ V[\underline{x}] = \underline{P}\,\underline{x} = [-1\ 0\ 1]^T$, shown in red in Figure 3.8.

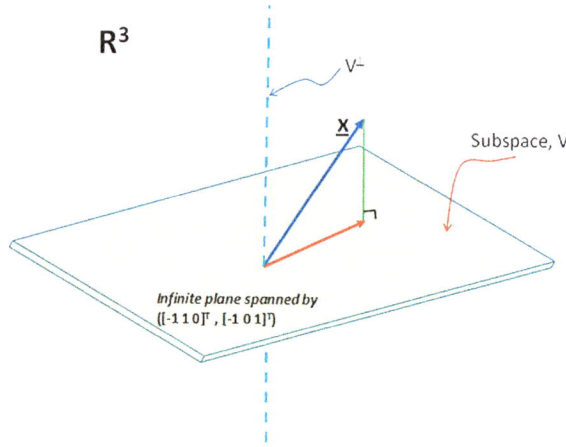

Figure 3.8 Subspace projection

A related and more interesting problem in preprocessing and ML feature extraction is the following.

$\underline{A}(\underline{x} = \underline{y}$. Given $\underline{x}$, find the $\underline{A}$ such that $\underline{y}$ has certain desired properties. A set of paired $\{\underline{x}, \underline{y}\}$ are provided as a training set. Using some method, such as an algorithm, find $\underline{A}$ such that the unwanted subset of $\underline{x}$ vectors lies in $N(\underline{A})$ and the wanted subset in the $R(\underline{A})$ with paired $\underline{y}$'s in $C(A)$.

The parallel to the digital filter design mentioned earlier is obvious. $\underline{x}$ vectors with high-frequency content will lie in $N(\underline{A})$; then the desired values for $\underline{y}$ become the low-pass output.

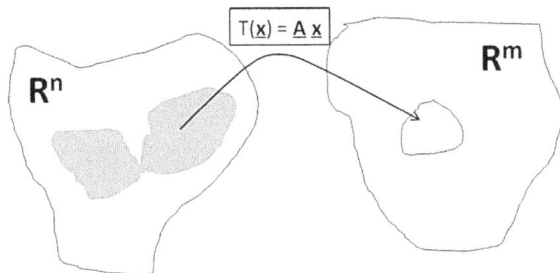

Figure 3.9 Subspace projection for filtering

Such an approach is commonly used in DSP when the input, x, is an additive combination of signal and noise. Given a desired signal, y, there are many powerful adaptive (learning) methods, all stemming from the Wiener optimum filter theory.

Orthonormal Basis

Often, we see a situation where $\underline{x}$ in standard coordinates (i.e., unit vector basis) requires a complicated transformation matrix, $\underline{A}$, to get a desired result (such as in computer graphics, where we want the linear transformation to yield a different perspective). In such cases, it is worthwhile to transform $\underline{x}$ from the standard coordinates to coordinates with respect to a new basis, such that the transformation operation is simplified. With this motivation, let us consider a general case.

If B is the basis for R^n, vector $\underline{x}$ can be represented using a different basis. This will result in different transformation, but the end-points will be the same.

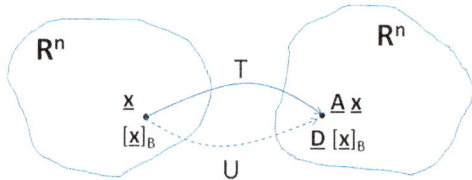

Figure 3.10 Choice of the basis

Let $B = \{v_1, v_2, \ldots, v_N\}$ be a basis for R^n. For the change of the basis matrix for

$$B, C = \begin{bmatrix} \uparrow & \uparrow & \uparrow & \uparrow \\ \mathbf{v}_1 & \mathbf{v}_2 & \cdots & \mathbf{v}_n \\ \downarrow & \downarrow & \downarrow & \downarrow \end{bmatrix}$$

i.e., $\underline{C}[\underline{x}]_B = \underline{x}$ or $[\underline{x}]_B = \underline{C}^{-1} \underline{x}$.

$[\underline{x}]_B$ is $\underline{x}$ represented in coordinates with respect to the B basis. Its position is unchanged in the domain. The transformation $T(\underline{x}) = A\underline{x}$ projects $\underline{x}$ into a specific point in the co-domain. Figure 3.10 shows that when we find the transformation matrix, D, using the new basis, the resulting end point should not be different. Figure 3.11 shows that $D = \underline{C}^{-1} A\underline{C}$.

If D has a simple form, subsequent operations can be performed in the coordinates with respect to B rather than in standard coordinates.

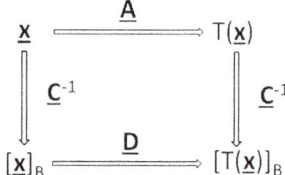

Figure 3.11 Basis transformation matrix

A Numerical Example

Let us look at a simple example. The linear transformation, $T(\underline{x}) = \underline{A}\underline{x}$.

Given $\underline{A} = \begin{bmatrix} 3 & 2 \\ 2 & -2 \end{bmatrix}$ and $\underline{x} = \begin{bmatrix} 1 \\ -1 \end{bmatrix}$. Alternate R^2 basis, $B = \left(\begin{bmatrix} 1 \\ 2 \end{bmatrix}, \begin{bmatrix} 2 \\ 1 \end{bmatrix} \right)$

For the change of the basis matrix for $B, \underline{C} = \begin{bmatrix} 1 & 2 \\ 2 & 1 \end{bmatrix}$.

You can easily get these results by using simple 2×2 matrix operations.

$$T(\underline{x}) = \underline{A}\underline{x} = \begin{bmatrix} 5 \\ 4 \end{bmatrix}$$

$$D = \underline{C}^{-1}\underline{A}\underline{C} = \begin{bmatrix} -1 & 0 \\ 0 & 2 \end{bmatrix}, \text{ a diagonal matrix.}$$

$$[\underline{x}]_B = \underline{C}^{-1}\underline{x} = -\frac{1}{3}\begin{bmatrix} 1 & -2 \\ -2 & 1 \end{bmatrix}\begin{bmatrix} 1 \\ -1 \end{bmatrix} = \begin{bmatrix} -1 \\ 1 \end{bmatrix}$$

$$\therefore [T(\underline{x})]_B = D[\underline{x}]_B = \begin{bmatrix} 1 \\ 2 \end{bmatrix}$$

Note that $T(\underline{x}) = \begin{bmatrix} 5 \\ 4 \end{bmatrix}$ and $[T(\underline{x})]_B = \begin{bmatrix} 1 \\ 2 \end{bmatrix}$ are the same vector in the co-domain. They are simply represented using a different basis.

The original problem is given in the top row of Figure 3.11 ($T(\underline{x}) = \underline{A}\underline{x}$). Using an alternate basis, B, we transformed the problem to the one in the bottom row, $[T(\underline{x})]_B = D[\underline{x}]_B$. If we compare the two transformations, we find that one involves a full matrix, $\underline{A}$, and the other a diagonal matrix, D. When there are hundreds or millions of rows, working in transformed coordinates using a change of basis matrix, $\underline{C}$, is significantly less work.

How do you know to choose the change of the basis matrix, $\underline{C}$, column vectors, v_i? Is there unique best choice?

We saw earlier that eigenvectors of $\underline{A}$ are special; $\underline{A}$ transforms them unchanged except for a scaling. Eigenvectors are orthogonal and hence form a linearly independent set of vectors. From prior linear algebra courses, you know that eigenvalues and eigenvectors are obtained by solving the following characteristic equation:

$|(\underline{A} - \lambda\underline{I}| = 0$, where $|.|$ is the determinant, $\{\lambda_i\}$'s are the N eigenvalues and $\{q_i\}$ are the N eigenvectors (we have tacitly assumed that $\underline{A}$ is a real, symmetric matrix).

Choose the change of basis matrix, $\underline{C} = \underline{Q} = \begin{bmatrix} \uparrow & \uparrow & \uparrow & \uparrow \\ q_1 & q_2 & \cdots & q_n \\ \downarrow & \downarrow & \downarrow & \downarrow \end{bmatrix}$

- The eigenvalues can be assembled into a diagonal matrix, $\underline{\Lambda} = \text{diag} (\lambda_1, \lambda_2, \ldots, \lambda_N)$.

Then, the transformation equivalent to $\underline{A}$ in the new eigenvector basis, $D = \underline{Q}^{-1} \underline{A} \underline{Q} = \underline{\Lambda}$, a diagonal matrix. Therefore, when you use the eigenvector basis in R^n, transformations become operations with diagonal matrices; nothing in linear algebra can be simpler than multiplying or inverting diagonal matrices.

Graphical Interpretation of Eigenvectors and Eigenvalues

To obtain a better understanding of the use of eigenvectors as a coordinate basis, let us look at random variables. In the last chapter, we mentioned the random time series and earlier in this chapter, we discussed ARMA models where the input to the LTI system is considered random variables or $\underline{x}$ as a random vector.

Let us look at the probability density functions (pdfs) of a random variable, in particular, Gaussian (or Normal) random variables. Figure 3.12 is the pdf of a single Gaussian random variable (often abbreviated as r.v.), the famous bell curve.

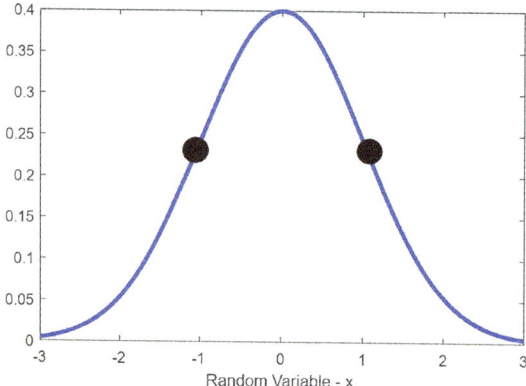

Figure 3.12 Gaussian probability density function

We want to investigate the unit standard deviation (usd) contour of the pdf. In this zero-mean, unit-variance univariate case, the "contour" is just the two points on the blue curve, one standard deviation away from the vertical through 0 (where pdf = 0.242). The joint pdf of the bivariate case is more interesting to visualize.

Just as in the univariate case, in the case of the "solid bell" for a bivariate Gaussian curve, the usd contour is obtained by passing a plane

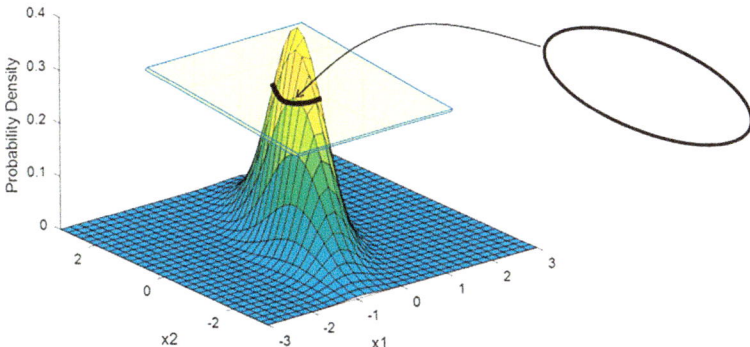

Figure 3.13 A bivariate Gaussian pdf

through where the standard deviation is 1. In general, the cross-section is an ellipse, as shown by simplifying the section into a circle when (x_1, x_2) are uncorrelated (Figure 3.13).

For the case of $\{x_n\}$ r.v.'s taken together, the joint pdf is hard to visualize (for $N = 3$, it is a hyper-ellipse or a "football" or a "watermelon" shape). The so-called correlation matrix captures the second-order statistics of the N-dimension joint pdf, which is what we need to visualize the usd contours.

The correlation matrix, $\underline{R} = E[\underline{x}\underline{x}^T]$, where $E[.]$ is the expectation operator. The n-dimensional multivariate Gaussian joint pdf is given by

$$f(x)_X = \frac{1}{2^{n/2} R^{1/2}} e^{-\frac{1}{2} x^T R^{-1} x}$$

$f(x)_X = 0.242$ when $\underline{x}^T \underline{R}^{-1} \underline{x} = 1$, the quadratic form that gives us the general equation of the unit standard deviation contour, which is a conic section. When we do a coordinate transformation using eigenvalue-vector decomposition that we saw in the last section, the quadratic form can be written in terms of eigenvalues and eigenvectors.

The eigenvalues and eigenvectors of $\underline{R}$ are as follows:

$$\Lambda = \text{diag}\,(\lambda_1, \lambda_2, \ldots, \lambda_N) \text{ and } \begin{bmatrix} \uparrow & \uparrow & \uparrow & \uparrow \\ q_1 & q_2 & \cdots & q_n \\ \downarrow & \downarrow & \downarrow & \downarrow \end{bmatrix} = \underline{Q}$$

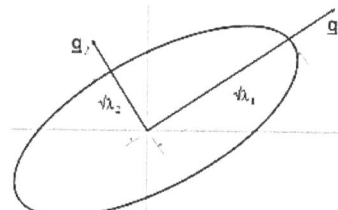

Figure 3.14 Unit standard deviation contour and eigenvectors and values

The eigenvectors point in the directions of the major and minor axes of the usd contour, and the length of the semi-major and minor axes are equal to the square roots of the corresponding eigenvalues (eigenvalues equal to the variances). The same insights hold for higher dimensions.

GRAPHICAL INTERPRETATION AND INSIGHTS FROM THE UNIT STANDARD DEVIATION CONTOURS

- In the latter portions of Part II of this book, we develop high-value features for ML applied to the hard problems of spatio-temporal data.

- Insights in this section will come in handy when developing algorithms using the Random Field theory.

CONCLUSION

In this chapter, we have reviewed a wide range of technical topics that were already familiar to you. What was gained was an understanding of the interconnections among them and some new insights, graphical and otherwise.

Linear algebra is the *lingua franca* of machine learning. We have done a review of linear algebra on an "as needed" basis in this chapter. We have also indicated how a few other disciplines, such as digital filtering and systems theory can be framed as linear algebra problems. The hope is that these interconnections will spark some ideas that will prompt you to add to the ML bag of tricks for your own technical domains.

4

"MODERN" MACHINE LEARNING

Many data scientists are self-taught. They have acquired various ML methods from the Web and online courses. Even some ML courses are taught as a collection of tools and tricks. This is typical of an emerging discipline, but ML is now sufficiently mature to be viewed from a unified perspective.

The advantages of a unified understanding are many. When we see ML within a common framework, the methods can be understood as variations rather than as entirely new techniques, which is key to learning any new discipline. Additionally, due to the ad hoc nature of ML development so far, there is a proliferation of terminology and explanations that can often be bewildering. So, in this chapter, we bring together what we reviewed in Chapters 2 and 3 and treat the basics of ML in a unified formal manner.

ML FORMALISM

There are many books that attempt to organize ML theory and practice, including one of the better options, a book by Simon Haykin [HS08]. Another good choice is a book by Abu-Mostafa [AY12]. In addition, we will borrow concepts from other sources to develop the unified approach below. Among the various ML learning systems, supervised learning provides a tractable case for formal learning theory analysis, shown as a system in Figure 4.1.

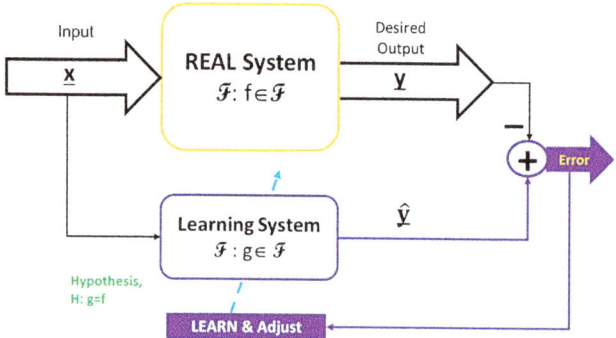

Figure 4.1 Learning theory as a system

Using training examples, the learning algorithm tries to "learn" the target function, f, to a good approximation, g. To be explicit,

- There is a set of learning system models.

- Each model in the set has a model structure and an associated learning algorithm.

- A model structure and associated learning algorithm is chosen. From the infinite set of functions, F, there is the TRUE target function, f, which is in unknowable and its good approximation, g, that we find using our choice of model structure, associated learning algorithm, and training data (Table 4.1).

Table 4.1 Learning models and algorithms

Model	Structure	Learning Algorithm	Objective
Perception	Perception nodes	Perception	
Logistic regression	Multiple regression	Gradient descent	Find g ≈ f
Neural network	Multiplayer perception	Back propagation	

When measuring the goodness of fit using unseen test data, g does not function as well as it did during training period, in general. This depends on the ability of g to generalize. Learning theory gives us some guidelines as to what a "good" approximation means, how well it generalizes, and how many training samples are required for sufficient learning. This is a difficult topic, but it is good to develop a general understanding.

Error Measures

For complex problems in classification where you are given N associated pairs of $\{\underline{x},\ \underline{y}\}$, we can make an assumption that even the "ideal" function, f, will not provide completely accurate classifications in a real-life hard multi-class problem. However, it is reasonable to assume that almost all values for $\underline{x}$ will be classified into the correct y classes. The function, g, that we learn can always be expected to perform even more poorly (other than in trivial cases).

Consider a two-class problem. Once we know the correct and incorrect classification produced (either during training or testing) on each $\underline{x}$, we can form the 2×2 table (Table 4.2).

Table 4.2 2x2 classification table

Classification Table:		f	
		Class#1	Class #2
g	Class #1	X	B
	Class #2	A	Y

The diagonal elements are the counts of the correct classification and the classification agreement between f and g. In a practical example, Class #1 may be "Terminal Disease" and Class #2 is "No Disease." B gives the proportion of false positives (Type I error) and A is the proportion of false negatives (Type II error).

The cost associated with false positives and negatives can be dramatically different in practical applications. In the case mentioned above, false positives may be tolerable since they mean only more testing and associated hardship and cost for the patient. However, a false negative in an ML system declaring that the patient does not have the disease when actually the patient does can be catastrophic. Errors that are tolerable and desirable are important user specifications that are application-dependent. This has to be an input to the learning system.

BAYES

We have implicitly considered the data to be deterministic so far. You are given a set of $\{\underline{x},\ \underline{y}\}$ and you learn a deterministic function, g, from them. Assume that y is a random variable affected by $\underline{x}$. Then instead of a function relationship, $\underline{x}$ and $\underline{y}$ are related by a conditional probability, $P(\underline{y} \mid \underline{x})$. You know the unconditional probability of x, $P(x)$ and the joint probability of x and y occurring, $P(x,y)$. The function learning problem in the deterministic case transforms into a conditional probability estimation problem in the random case.

Consider a two-class problem where certain values of y indicate Class #1 and others, Class #2. Then, given $\underline{x}$, what is the conditional probability, $P(\text{Class \#1} \mid \underline{x})$ and $P(\text{Class \#2} \mid \underline{x})$. If the first probability is higher, we say $\underline{x}$ belongs to Class #1; if not, Class #2. This is the Bayes Classifier.

Using the joint density definition, we can write the following:

$P(\text{Class \#1} \mid \underline{x}) = P(\text{Class \#1}, \underline{x})/P(\underline{x})$.

Also, $P(\underline{x} \mid \text{Class \#1}) = P(\text{Class \#1}, \underline{x}) / P(\text{Class \#1})$.

$$\therefore P(\text{Class \#1} \mid \underline{x}) = \frac{P(x \mid \text{Class \#1})}{P(x)} * P(\text{Class \#1}).$$

This is the Bayes Theorem.

Classification Rule: $\underline{x} \in$ Class #1 iff $P(\text{Class \#1} \mid \underline{x}) > P(\text{Class \#2} \mid \underline{x})$

This classification rule (called the "maximum a posteriori" or "MAP" rule) takes the place of finding function, g, in the deterministic case.

We can re-write the classification rule as follows:

$$\underline{x} \in \text{Class \#1 iff } \frac{P(x \mid \text{Class \#1})}{P(x)} * P(\text{Class \#1}) > \frac{P(x \mid \text{Class \#2})}{P(x)} * P(\text{Class \#2}).$$

$\underline{x} \in$ Class #1 iff $P(\underline{x} \mid \text{Class\#1})*P(\text{Class \#1}) > P(\underline{x} \mid \text{Class \#2})*P(\text{Class \#2})$

$$\underline{x} \in \text{Class \#1 iff } \frac{P(x \mid \text{Class \#1})}{P(x \mid \text{Class \#2})} > \frac{P(\text{Class \#2})}{P(\text{Class \#1})}$$

$\frac{P(x \mid \text{Class \#1})}{P(x \mid \text{Class \#2})} = L(\underline{x})$, called the "likelihood ratio." $\frac{P(\text{Class \#2})}{P(\text{Class \#1})} = \theta$, called a "threshold."

$$\therefore \ \underline{x} \in \text{Class \#1 iff } L(\underline{x}) > \theta; \underline{x} \in \text{Class \#2 iff } L(\underline{x}) < \theta.$$

Note that the likelihood expression is a function of the same form as function, g, in the deterministic case.

The ROC (Receiver Operating Characteristics) Curve

As you vary θ in the likelihood classifier, the classification error will vary. In the 2×2 table in the previous section, entries will shift around.

True Positive Rate $= X_{\text{,}} (X + B)$

False Positive Rate $= B_{\text{,}} (X + B)$

These two values are plotted for various values of θ in Figure 4.2. When done for three different classifiers, it is clear why the diagonal line is a worthless classifier (just as good as tossing a coin). The "Excellent" one (shown in yellow) is closer to the "knee" of the curve and produces a high (around 0.9) probability of true positives with only a low 0.1 probability of false positives.

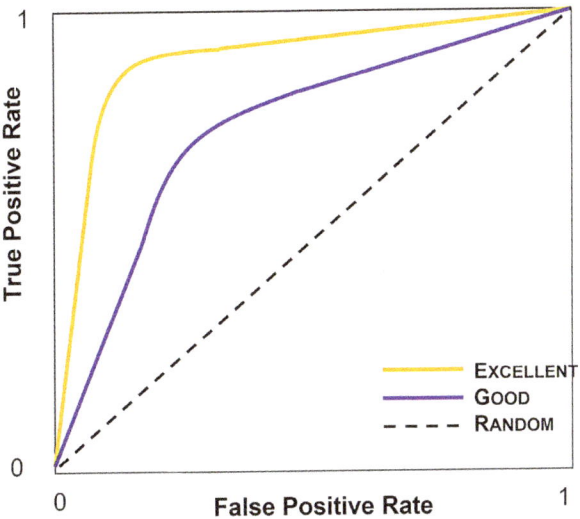

Figure 4.2 Receiver operating characteristic curve

ROC curves are valuable in evaluating and comparing the performance of two-class classifiers. Unfortunately, extensions to multiple classes are not easy.

GENERALIZATION, THE HOEFFDING INEQUALITY, AND VC DIMENSION

A challenging problem in machine learning (and statistics and systems theory) is the model order selection. In ML, it appears as the difficulty in assuring that a good approximation function, g, learned during training also performs well during the test. This is the *generalization problem.*

Consider the Training Set as a random sample of all possible values for $\underline{x}$. We want to estimate a population parameter, μ, by a statistic, ρ, measured from a sample of some size, N. The Hoeffding inequality states the following probability relationship:

$$P[|r - \mu| > \varepsilon] \leq 2 \exp(-2\,\varepsilon^2\,N) \text{ for any } \varepsilon > 0.$$

As N becomes large, it becomes exponentially unlikely that r will deviate from μ by greater than ε. Based on some rigorous development, one can extend the inequality to the cases of Training Error, E_{TR}, and Test Error, E_{TS}.

$$P[|E_{TR} - E_{TS}| > \varepsilon] \le 2 \exp(-2\,\varepsilon^2\,N).$$

In other words, as we have more Training data, N, the Training and Test errors will be closer, which means the generalization will be better in that performance under the Training and the Test results will similar. (However, this says nothing about how good the performance itself is.)

The tradeoff is between good generalization (low $|E_{TR} - E_{TS}|$) and acceptable classification error (low $|E_{TR} + E_{TS}|$). By allowing our function, g, to be more complex, we can fit the Training Set better, but the ability to generalize may become poorer in the extension. We need to re-write the Hoeffding Inequality.

$$E_{TS} \le E_{TR} + \sqrt{\frac{1}{2N} \ln \frac{2M}{\delta}}$$

With a probability of $(1 - \delta)$ or better, the Test Error will be close to Training Error within the 2nd term. Notice a new variable, M, which brings complexity into the equation and helps with the tradeoff. We need more complexity in function g to reduce both errors. However, too much complexity in g will reduce the generalization's ability.

A larger value of M will cause the Test Error bound to increase. As in curve fitting, a very high order polynomial will "curve" through every data point in the Training Set, but will leave many other new data points outside the fitted curve. A higher value of N will reduce the 2nd term and drive the two errors closer, indicating a better generalization. The desirable M is small and N must be large to meet the trade-off.

M comes from a much deeper mathematical consideration called the Vapnik-Chervonenkis (VC) dimension. Let us say that we are given a Training Set of data points. A greater number of distinct points creates more combinations of labeling and therefore tests a greater number of potential decision boundaries. Correctly classifying all the labeling combinations means that the classifier is capable of forming all of these potential decision boundary configurations. Suppose we are able to do that with a maximum of M data points. Then, the VC dimension is M.

In the theory of learning, the last equation provides the relationship between the errors and number of training data points and the complexity of the mapping functions and generalizability.

FORMAL LEARNING METHODS

There is a unified perspective to gather various supervised learning methods in ML and a few other extensions. As we have done all throughout this book, we will assume a knowledge of basic ML and not develop results from first principles.

The objective of any learning method is to minimize the total error function, E_{tot}:

- Error at any instant, 'n' $= e[n] == y[n] - \hat{y}[n]$

- $E[n] = \frac{1}{2} e^2[n]$; then, $E_{tot} = \sum_{\text{All } n} E[n]$

Using the familiar method of steepest descent, first we find the gradient with respect to the node weights that determine the function, g.

$$\nabla_w E_{tot} = \frac{\partial E_{tot}}{\partial w} = \sum_{\text{All } n} \frac{\partial E_{tot}}{\partial w} = \sum_{\text{All } n} \nabla_w E[n]$$

This is appropriate for batch mode operations, but for real-time updates and considering scalar output, we have to use the *instantaneous estimate*, $\nabla_w E[n]$, instead of $\nabla_w E_{tot}$.

$$\frac{\partial E[n]}{\partial w[n]} = \frac{\partial}{\partial w} = \left(\frac{1}{2} e^2[n] \right) = - e[n] \frac{\partial \hat{y}[n]}{\partial w[n]} \qquad (1)$$

since desired output, $y[n]$, is not dependent on weights.

A change to any weight can minimize the error:

$$\Delta w[n] = \eta \frac{\partial E[n]}{\partial w[n]} = + \eta e[n] \frac{\partial \hat{y}[n]}{\partial w[n]} \text{ where } h \text{ is the "step-size"} \qquad (2)$$

This weight update rule is the basis of back propagation in a neural network where the derivatives are back propagated via the chain rule from the output, $\hat{y}[n]$.

Let us generalize the learning defined by Equation (2).

Given N training data samples, "cost function" or average error energy (as a function of weights) that we want to minimize by choosing function g appropriately is as follows:

$$E_{av}(\underline{w}) = \frac{1}{N}\sum_{n=1}^{N}E[n] = \frac{1}{2N}\sum_{n=1}^{N}e^2[n]$$

$E_{av}(\underline{w})$ can be expanded in a Taylor Series around the current operating point, $\underline{w}[n]$.

$$E_{av}(\underline{w}[n] + \Delta\underline{w}[n]) = E_{av}(\underline{w}[n]) + \underline{g}^T[n]\,\Delta\underline{w}[n] +$$

$$\tfrac{1}{2}\,\Delta^T\underline{w}[n]\,\underline{H}[n]\,\Delta\underline{w}[n] + \text{higher-order terms.} \qquad (3)$$

where $\underline{g}[n]$ is the local gradient vector of the error surface and the matrix, called *Hessian*, $\underline{H}[n]$, is the curvature of the error surface. A collection of $\underline{g}[n]$ for all N samples is the matrix called *Jacobian*, $\underline{J}$.

The combined use of Jacobian and Hessian (gradient and curvature) matrices can give excellent learning speed and a globally optimal solution for the weight vector if the error surface is not too complex.

The instantaneous estimate of the weight change in Equation (2) is a special case of Equation (3), where only the linear approximation is taken into account.

$$\Delta\underline{w}[n] = -\eta\,\underline{g}[n]$$

To improve the convergence performance, we can use the Hessian matrix.

$$\Delta\underline{w}[n] = \underline{H}^{-1}[n]\,\underline{g}[n]$$

These updates correspond to many well-known and powerful techniques that can be used for machine learning.

Table 4.3 Optimization methods for machine learning

Optimization Method	Derivative	Features	Remarks
(Stochastic) Gradient Descent	Jacobian	Easy implementation; slow convergence.	Most ML learning, including neural networks.
Newton	Hessian	Calculation of $\underline{H}^{-1}$; fast convergence.	$\underline{H}$ can be singular. Heavy computation & storage.

Optimization Method	Derivative	Features	Remarks
Quasi-Newton	Jacobian	Modified $\underline{H}$ to be positive-definite used.	
Conjugate-gradient	Quadratic	$\underline{H}$ not used directly.	
Marquardt	Hessian	$\underline{H}$ regularized.	Suited for non-linear least squares.
Dynamic Programing	Bellman	Current learning considers future goals.	For reinforcement learning.

Dynamic programing does not belong directly in this table but reinforcement learning can be "over-laid" on supervised learning to make "smarter" local decisions. In other words, let potential longer-term wins counter-weigh current (temporary) loss so that overall result is optimum. This is a topic for future development.

REGULARIZATION & RECURSIVE LEAST SQUARES

Learning from examples as we do in ML is called an *inverse problem* in that the direct problem governed by the physical laws that generated these examples and we are trying to reverse engineer the governing laws using ML.

ML is a collection of methods to construct a hypersurface or a multi-dimensional mapping that defines the output patterns in terms of input patterns. Since the hypersurface may not be smooth (and hence, learning from samples may lead to overfitting with poor eventual generalization performance), it is desirable to limit the hypersurfaces to a subset of smooth ones. Tikhonov's regularization theory provides the theoretical framework for smoothing.

We will review regularization in a particular context rather than address the general theory. The recursive least squares approach is a good learning method to consider.

As we saw in the last section, we attempted to minimize the Total Error, E_{tot} or its approximations, by developing a step-by-step weight change procedure. In our recursive least squares (RLS) formulation, we can minimize a modified error:

$$E_{RLS}(w) = \frac{1}{N}\sum_{n=1}^{N}E[n] = \frac{1}{2N}\sum_{n=1}^{N}e^2[n] + \frac{1}{2}\lambda\,|\underline{w}|^2 \qquad (4)$$

The last term is called a *regularizer* and λ is the *regularization parameter*. From the ordinary batch least squares solution when the minimized error does not have the last term in Equation (3) (we mentioned this in Chapter 3), the least squares solution for $\underline{w}$, $\underline{w}* = (\underline{ATA})^{-1}\,\underline{AT}\;\underline{y}$. If we re-write for the diagram in the last section, $\underline{y} = \underline{w}^T\,\underline{x}$, $\underline{y}$ is the output, $\underline{x}$ is the input, and the values for $\underline{w}$ are the weights (or coefficients) of the multiple linear regression equation. With appropriate vector dimensions, we find

$$\therefore \qquad \underline{w}* = (\underline{x}^T\underline{x})^{-1}\underline{x}^T\;\underline{y} = \underline{R}^{-1}\underline{r},\;\underline{R}$$

is the autocorrelation matrix of input, $\underline{x}$, and $\underline{r}$ is the cross-correlation vector between input, $\underline{x}$, and desired output, $\underline{y}$. This is the Normal equation.

When we minimize the RLS Error as in Equation (3), the least squares solution for the weights is as follows:

$$\underline{w}_R* = (\underline{R} + \lambda\underline{I}\;)^{-1}\;\underline{r} \qquad (5)$$

where $\underline{I}$ is the identity matrix of the proper order.

By setting different values for λ (from 0 to ∞), various levels of smoothness can be obtained for the resulting hypersurface, thereby providing a better generalization for the Test Set without sacrificing fidelity to (or full learning from) the Training Set.

Recursive Least Squares (RLS)

RLS is a well-known algorithm [HS08]. As we see in the Normal equation and Equation (4), the inversion of the autocorrelation matrix is a required step. Since this is computationally heavy, we want to work around it, which leads to the recursive solution. The burden of inverting $\underline{R}$ at each time step is avoided by the use of the matrix inversion lemma (used below) which underpins many powerful recursive estimation algorithms.

For the RLS algorithm,

Call $\underline{P} = \underline{R}^{-1}$

Given $\{\underline{x}[n],\;\underline{y}[n]\}$ for $n = 1$ to N and with $\underline{w}[0] = \underline{0}$ and $\underline{P}[0] = \lambda^{-1}\underline{I}$ with $\lambda =$ small positive constant, perform the following updates for each recursive step, n.

$$\underline{P}[n] = \underline{P}[n-1] - \frac{P[n-1]x[n]x^T[n]P[n-1]}{1 + x^T[n]P[n-1]x[n]}$$

$$\underline{g}[n] = \underline{P}[n]\,\underline{x}[n]$$

$$\alpha[n] = y[n] - \underline{w}^T[n-1]\,\underline{x}[n]$$

$$\underline{w}[n] = \underline{w}[n-1] + \underline{g}[n]\,\alpha[n]$$

Note that the RLS solution and maximum a posteriori probability (MAP) solution in the random variable case are the same, whereas the maximum likelihood solution is the same as the Normal equation (which is not a regularized solution). We know that maximum likelihood solution is unbiased, but of a high variance. The MAP solution, however, is biased but of a lower variance. The RLS solution can be considered to be a low variance but biased least squares solution.

REVISITING THE IRIS PROBLEM

In Chapter 2, we discussed classification of the Fisher Iris data. Here, we revisit using the formalisms we covered earlier in this chapter and Chapter 3. The Iris dataset has four attributes (sepal length, sepal width, petal length, and petal width) and three labelled classes (Iris: setosa, versicolor and virginica). To enable the visualization of results, we drop the last attribute and work with three attributes and three classes, as shown in Figure 4.3.

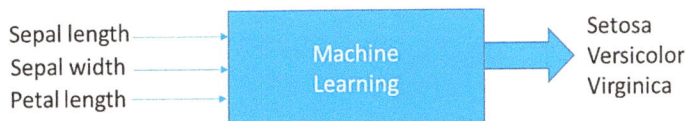

Figure 4.3 A visual representation of the three attributes and classes

Compare Figure 4.3 to our standard linear algebra and systems notation.

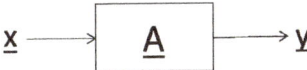

Attribute vector, $\underline{x}$, is a (3×1) vector and $\underline{y}$ is a (3×1) vector. $\underline{A}$ is a (3×3) matrix and $\underline{y} = \underline{A}\,\underline{x}$

To demonstrate the use of the linear algebra techniques we have reviewed in Chapter 3, we follow these steps:

1. Normalize the attributes.

2. Visualize them using a scatter diagram of any two attributes.

3. Convert the attributes to features:

- Find a suitable subspace.

- Find an orthonormal basis set of vectors for the subspace.

- Project attributes onto the subspace to obtain features.

4. Visualize the features and compare them to the attributes.

5. Split the data into the Training and Test sets.

6. Using the Training Set, develop a classifier.

7. Measure the Training Error.

8. Use the classifier to classify the Test Set.

9. Measure the Test Error.

10. Analyze the generalization using the VC dimension.

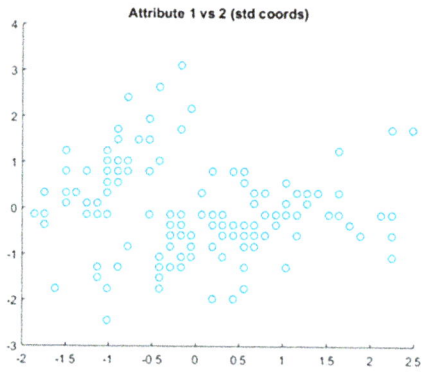

Figure 4.4 Scatter plots of attributes

Scatter Diagram of Attributes

The attributes, $\underline{x}$, are somewhat widely spread, which makes the classifier's job challenging. However, there is a separation between the cluster of points towards the upper left side of the plot and the rest of the points. Since we know there are three classes, separating them fully will be a challenging task since we cannot do it simply by looking at it.

Suitable Subspace: One of the well-established approaches to finding a subspace and its basis vectors is via eigen-decomposition.

Collecting all the 150 attributes into a $(3 \times N)$, where $N = 150$, matrix, $\underline{X}$, yields

$$X = \begin{bmatrix} x_{11} & x_{12} & \cdots x_{1,150} \\ x_{21} & x_{22} & \cdots x_{2,150} \\ x_{3,1} & x_{3,2} & \cdots x_{3,150} \end{bmatrix}$$

For our data, the (3×3) correlation matrix,

$$R = \begin{bmatrix} 0.9933 & -0.1086 & 0.8659 \\ -0.1086 & 0.9933 & -0.4177 \\ 0.8659 & -0.4177 & 0.9933 \end{bmatrix}$$

When we perform eigen-decomposition of $\underline{R}$, we find the eigenvalues to be $\{2.0004, 0.9087, 0.0709\}$. The first two eigenvalues are significant and the 3rd one is negligibly small.

These are the three eigenvectors of $\underline{R}$. So, a suitable subspace to consider will be the one spanned by the first two eigenvectors.

$$\begin{bmatrix} \phi_1 & \phi_2 & \phi_3 \\ 0.6314 & 0.4277 & -0.6469 \\ -0.3542 & 0.9011 & 0.2501 \\ 0.6898 & 0.0713 & 07205 \end{bmatrix}$$

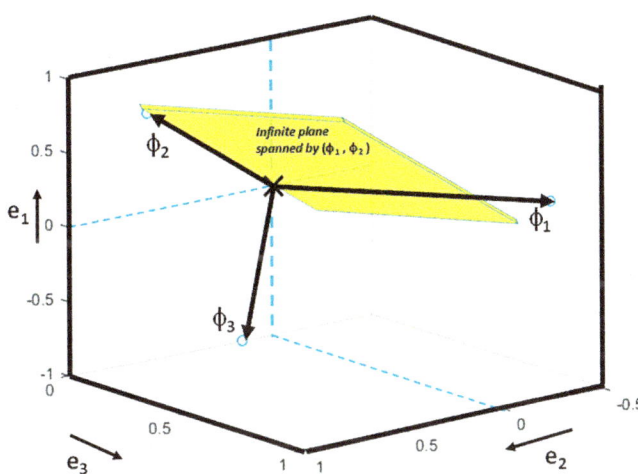

Figure 4.5 A three-dimensional visualization of the attributes

The subspace of interest to us is shown in yellow. Visualize this infinite plane as a "flat roof" coming out of the page at you, titled up slightly.

To obtain our features, we project each of the 150 Attribute vectors onto this yellow subspace. Call this section *subspace E*. *E* is a subspace of R^3 and $\{\phi_1, \phi_2\}$ are the two basis vectors for *E*. We construct a matrix, $\underline{C}$ such that its column vectors are $\{\phi_1, \phi_2\}$.

$$C = [\phi_1 \phi_2] = \begin{bmatrix} 0.6314 & 0.4277 \\ -0.3542 & 0.9011 \\ 0.6898 & 0.0713 \end{bmatrix}$$

The projection of the attribute vector, $\underline{x}$, is $\text{Proj}_E[\underline{x}] = \underline{C}(\underline{C}^T \underline{C})^{-1} \underline{C}^T \underline{x}$, or the projection matrix, $\underline{P} = \underline{C}(\underline{C}^T \underline{C})^{-1} \underline{C}^T$.

$$\text{Using the } \underline{C} \text{ matrix above, } P = \begin{bmatrix} 0.5816 & 0.1617 & 0.4660 \\ 0.1617 & 0.9375 & -0.1802 \\ 0.4660 & -0.1802 & 0.4810 \end{bmatrix}$$

Our new feature vector, $\underline{f} = \text{Proj}_E[\underline{x}] = \underline{P} * \underline{x}$.

We compare the scatter plots of attributes, $\underline{x}$, and features, $\underline{f}$, in the two plots in Figure 4.6.

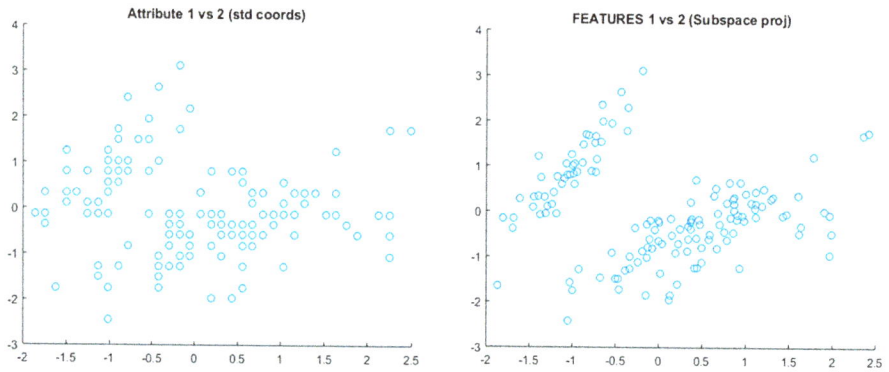

Figure 4.6 Scatter plots of attributes and features

Clearly, the feature clusters seem to be more tightly packed, but we do not see three well-separated clusters. That is the job of the classifier.

As we had observed in Chapter 3, the features, their probability density functions, and unit standard deviation (USD) contours are very insightful. If Feature 1 and Feature 2 were bivariate Gaussian, the usd contour will appear as shown in Figure 4.7.

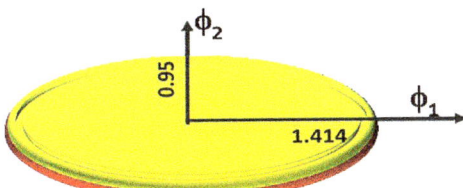

Figure 4.7 The usd contour when Feature 1 and Feature 2 are bivariate Gaussian

The yellow elliptical disc lies on the same plane as the yellow "roof" in the 3D picture, which is the plane of ϕ_1 and ϕ_2. These two eigenvectors are orthogonal from the top view, which is not obvious in the 3D picture. The semi- major and minor axes lengths are equal to the square roots of the two largest eigenvalues (as we have observed in Chapter 3).

Classifier

We introduced the RSL algorithm in the last section, which is a regularized solution with a low variance (but with some bias). We apply the RLS to the Iris features in this section to solve the classification problem and compare to our solutions to those found in Chapter 2.

With a choice of $\lambda = 0.1$ and the features in the Training Set randomized, the converged Training MSE = 1.9066. As you can see in Figure 4.7, the convergence of RLS is very rapid, and occurs after fewer than 10 training samples.

Figure 4.8 Training error plot for the RLS algorithm

The Test Set classification performance is shown in Figure 4.9. As we did previously, the correct classification labels are shown in red (Test MSE = 4.2709).

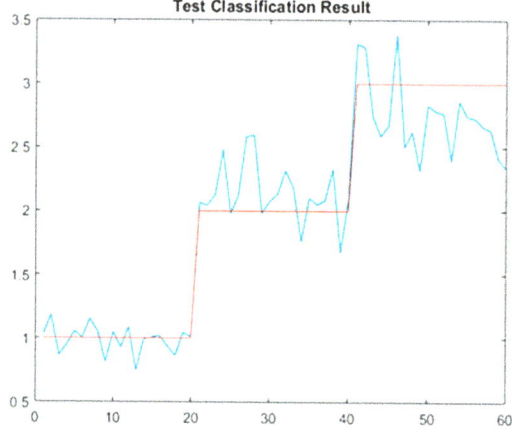

Figure 4.9 Test set classification results

From our previous discussion of the VC Dimension and generalization, we expect the Test Error to be bounded as follows:

$$E_{TS} \le E_{TR} + \sqrt{\frac{1}{2N} \ln \frac{2M}{\delta}}$$

Unfortunately, since the VC Dimension, M, for this Iris Training set is not known, we can only write the following:

$$4.2709 \le 1.9066 + \sqrt{\frac{1}{180} \ln \frac{2M}{0.05}} \quad \text{for a 95\% probability bound.}$$

For the upper bound when it is an equality, the square root term = 2.3643.

KERNEL METHODS: NONLINEAR REGRESSION, BAYESIAN LEARNING, AND KERNEL REGRESSION

In the last section, RLS provided us fast convergence and regularization to control generalization (smoothing) performance. But the RLS results in hyperplanes (linear hyper-surfaces) that separate the different classes. What about classes that are only separable non-linearly?

We can use backpropagation neural networks; they are well-known and we will not discuss them further in this chapter other than to mention that they can be very good solutions to many ML problems.

In this section, we bring together nonlinear regression, conditional expectation, Bayesian learning, Parzen-window density estimation, the kernel regression estimator, and radial basis function networks and apply them to the Iris problem of the last section.

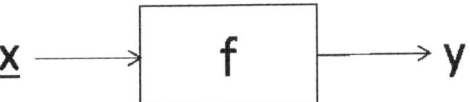

Given N pairs of $\{x, y\}$, where the values for x are the input features and the values of y are the desired output in the Training Set, our problem is to find f, where f is a nonlinear function.

$$y_i = f(x_i) + e_i \text{ where } i = 1, 2, \ldots, N \tag{1}$$

Here, e_i is an additive white noise with a zero mean and variance $= s_e^2$.

The unknown function, f, is the conditional expectation (mean) of y given the regressor, x.

$$f(x) = E[y|x] = \int_{-\infty}^{\infty} y\, p_{y|x}(y|x)\, dy \tag{2}$$

From joint probability density definition, we obtain

$$p_{y|x}(y|x) = \frac{p_{x,y}(x,y)}{p_x(x)} \tag{3}$$

Once we have learned the a-posteriori density of y, $p_{y|x}(y|x)$, the conditional expectation calculated using Equation (2) gives us $f(x)$. But first, we have to estimate the joint density function, $p_{x,y}(x,y)$, from which the marginal density, $p_x(x)$ can be obtained easily by integrating out y.

To estimate the joint density from the given data, we use the Parzen-window estimator. Parzen window or "kernel" can take many forms. A function $k(x)$ is a kernel as long as (1) it satisfies certain continuity requirements and (2) the volume under the kernel is unity. The Gaussian function is a popular kernel.

The attractiveness of what we have developed so far becomes clear in the light of the cover theorem, which states that "when a classification problem is transformed non-linearly into a high dimensional space, the classes are likely to be linearly separable." On this basis, we proceed with the non-linear regression approach.

From the given data, to estimate joint density and the marginal density functions that we require, we can select any Parzen window (rectangle, triangle, exponential, and the most popular, Gaussian). The kernel for the Parzen Gaussian window is as follows:

$$k\left(\frac{x-x_i}{h}\right) = \frac{1}{h^m \sqrt{2\pi}} \exp{-\frac{1}{2}\left(\frac{x-x_i}{h}\right)^2} \tag{4}$$

where $\underline{x}$ is the input vector of dimension, m, from the Training Set and h is a selectable positive scalar called the width; in other words, this is a bell-shaped surface, centered at $\underline{x}_i$ with a uniform width, h. In essence, we are trying to identify a set of "bells" that, when added together, will approximate the nonlinear hypersurface that separates the classes.

We use the Parzen window method in Equation (4) to find the joint and marginal densities, $p_{x,y}(x,y)$ and $p_x(x)$ in Equation (3) and perform the integration in Equation (2) to get an estimate of $f(\underline{x})$.

$$\hat{f}(x) = \frac{\sum_{i=1}^{N} y_i \, k\left(\frac{x-x_i}{h}\right)}{\sum_{j=1}^{N} k\left(\frac{x-x_j}{h}\right)} = \sum_{i=1}^{N}(x,x_i)y_i \quad \text{with} \quad \psi_i(x,x_i) = \frac{k\left(\frac{x-x_j}{h}\right)}{\sum_{j=1}^{N} k\left(\frac{x-x_j}{h}\right)},$$

for $i = 1, 2, \ldots, N$.

From a regression model perspective, we non-linearly transform the input, $\underline{x}$, multiply by a weight, and sum them to get the regression output. Therefore, we notice in the above equation that y_i plays the role of regression coefficients or weights, $w_i = y_i$. Then,

$$\hat{f}(x) = \sum_{i=1}^{N} w_i \, \psi_i(x,x_i) \text{ for } i = 1, 2, \ldots, N. \tag{5}$$

For any Test Set input, $\underline{x}$, use this $\underline{x}$ against all N Training Set data, $\{\underline{x}, y\}$; first perform Parzen window kernel evaluation and then scale by weight, $w_i = y_i$. Obtain the sum over all the Training Set data to get the regressor/classifier output.

$\hat{f}(x)$ in Equation (5) is the basis for many well-known ML techniques (such as the Naïve Bayes, radial basis function network, kernel regression, and support vector machine). Note that Equation (5) is not the solution to all problems, though. For example, there are as many kernels as there are training samples: N numbers of $\underline{x}_i$. A large training set size can be a good thing, but the computational burden of Equation (5) becomes large. If you reduce the number of kernels from N, the question of where to center the kernel arises (k-means clustering can provide the centers).

Equation (5) and the steps leading to it embody all the critical concepts in ML (the cover theorem, nonlinear regression, conditional expectation, Bayesian learning, kernel regression, and the basis for support vector machines).

A Numerical Example

Continuing our Iris example with only three features from the previous section, we now implement the classifier in Equation (5). Note that this is not a recursive or step-by-step learning method, but a block method. Here, we are interested in the non-linear classifier performance.

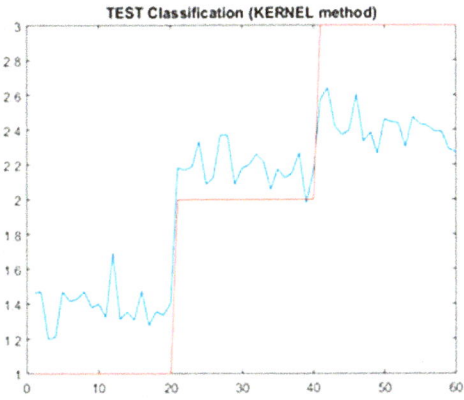

Figure 4.10 Classification results for kernel method

With a choice of $h = 1.5$ (a bit more smoothing than when $= 1$), we get the Test results shown in Figure 4.10. Test MSE = 0.1860. Note that this is significantly better than the Test MSE = 4.2709 that we obtained for the linear RLS case in the last section.

The MATLAB m code for the kernel method is part of the RLS code in the chapter appendix.

RANDOM PROJECTION MACHINE LEARNING

The authors of this method call it "extreme" machine learning [HG12]. There are some controversies regarding the originality of the formulation, but Random Projection Machine Learning (RPML) has some useful properties, such as a lower number of hidden nodes than the kernel method of Equation (5) in the last section.

RPML is based on the cover theorem, whereby input is non-linearly projected onto a high-dimensional space to solve the classification problem. Instead of carefully chosen and adaptively adjusted Gaussian (or other) kernels, RPML uses a small set of random Gaussians (or other non-linearities, even a mixture of them); the linear weights of the output node are solved by (regularized) least squares such that some of these random Gaussians make up the non-linear hypersurface. This low-complexity method works well for many classification problems.

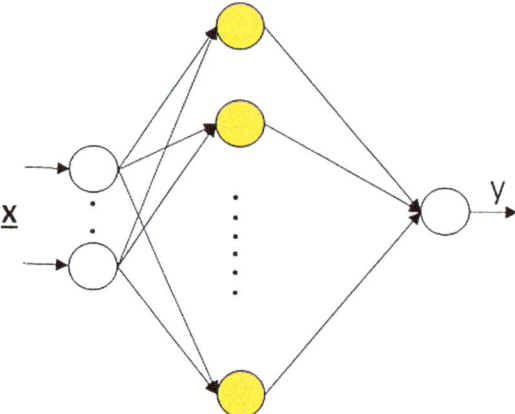

Figure 4.11 Random projection machine learning structure

RPML has a single hidden layer (yellow) and nodes with nonlinear activation functions, as shown in Figure 4.11. The weights connecting the input nodes to the hidden layer nodes are random (assigned at the outset) and they are not updated. The linear weights connecting the hidden layer nodes to the output node is estimated by the least squares. RPML has very simple dynamics and updates.

The hidden node, m, output can be written as follows:

$$s_m[n] = \underline{w}_m^{i\,T} \, \underline{x}[n] + b_m \text{ for } n = 1 \text{ to } N$$

where N is the Training Set size. $\underline{w}_m^i$ are the values for the random weights from the input nodes to the hidden nodes; the values for b_m are the random bias terms at each hidden node.

$$y[n] = \sum_{m=1}^{M} w_m^o \rho(s_m[n]) \text{ for } n = 1 \text{ to } N. \tag{6}$$

where M is the number of hidden nodes (usually $< N$, unlike in the kernel method in the last section). $\rho(.)$ is a nonlinear function such as Gaussian (ρ for all hidden nodes need not be the same type but here we are choosing the same nonlinearity). w_m^o is the weight between a hidden node, m, and the output node, which is to be solved for the Training Set desired outputs for $n = 1$ to N.

The N equations in Equation (6) can be written in matrix form using an $(N \times M)$ data matrix, $\underline{D}$:

$$\underline{D} = \begin{bmatrix} \rho(w_1^{iT} x[1] + b_1) & - & \rho(w_M^{iT} x[1] + b_M) \\ - & - & - \\ \rho(w_1^{iT} x[N] + b_1) & - & \rho(w_M^{iT} x[N] + b_M) \end{bmatrix}$$

The class labels in the Training Set form the desired response vector of size $(N \times 1)$, $\underline{d}$.

$\underline{d} = \underline{D} * \underline{w}^o$ where $\underline{w}^o$ is the $(M \times 1)$ output node weight vector.

The least squares estimate of the output weights,

$$\widehat{w^o} = (\underline{D}^T \underline{D})^{-1} \underline{D}^T \underline{d} \tag{7}$$

Steps in the RPML Algorithm:

1. Choose M, the number of hidden nodes.

2. Assign random numbers to all $\underline{w}_m^i$ and b_m.

3. Calculate $\underline{D}$ for all Training Set inputs.

4. Using the Training Set class labels, use Equation (7) to solve for the output weights, $\widehat{w^o}$.

5. Use $\widehat{w^o}$ with the Test Set features transformed by $\rho(.)$ to predict the classes.

For example, let's continue with our Iris problem with three features from the previous section and implement RPML. Note that this is not a recursive or step-by-step learning method but a block method.

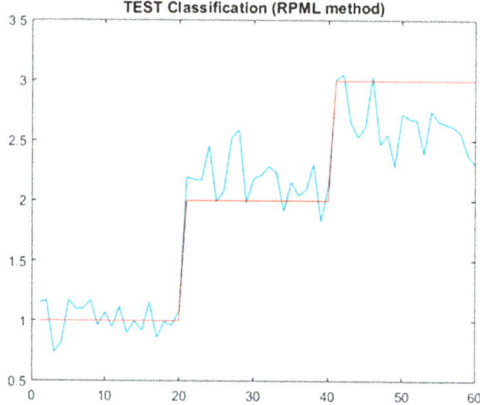

Figure 4.12 Classification results using the RPML method

With 10 hidden nodes and sigmoidal non-linearity (simpler than Gaussian), the Test MSE = 0.0886, which is lower than in the last section's kernel method!

RANDOM PROJECTION RECURSIVE LEAST SQUARES (RP-RLS)

The D matrix in the RPML method can be processed recursively at each time instant or per each Training Set feature.

$$\underline{D} = \begin{bmatrix} \rho(w_1^{iT} x[1] + b_1) & - & \rho(w_M^{i\,T} x[1] + b_M) \\ - & - & - \\ \rho(w_1^{iT} x[N] + b_1) & - & \rho(w_M^{i\,T} x[N] + b_M) \end{bmatrix}$$

This is the RLS algorithm we discussed earlier, which also has the regularization control to impact generalization (or the smoothness of the mapping function).

For the RLS algorithm, do the following:

Call $\underline{P} = \underline{R}^{-1}$

Given $\{\underline{x}[n],\ \underline{y}[n]\}$ for $n = 1$ to N and with $\underline{w}[0] = 0$ and $\underline{P}[0] = l^{-1}\underline{I}$ with λ = small positive constant, perform the following updates for each recursive step, n.

$$P[n] = P[n-1] - \frac{P[n-1]x[n]x^T[n]P[n-1]}{1+x^T[n]P[n-1]x[n]}$$

$$g[n] = P[n]\,\underline{x}[n]$$

$$\alpha[n] = y[n] - \underline{w}^T[n-1]\,\underline{x}[n]$$

$$\underline{w}[n] = \underline{w}[n-1] + g[n]\,\alpha[n]$$

In the RP-RLS,

$$\underline{x}^T[n] = \left[\rho\left(r_1^{iT}\,x[n]+b_1\right)\rho\left(r_2^{iT}\,x[n]+b_2\right)\cdots\rho(r_M^{i\ T}\,x[n]+b_M)\right]$$

In the RPML equation, $s_m[n] = \underline{r}_m^{i\ T}\,\underline{x}[n] + b_m$ is re-written with $\underline{r}$ instead of $\underline{w}$ to avoid confusion with the RLS weights, $\underline{w}$. M is the number of hidden nodes.

For the RP-RLS algorithm, do the following:

Given $\{\underline{x}[n], y[n]\}$ for $n = 1$ to N.

Choose M and function $\rho(.)$.

Assign random numbers to all $\underline{r}_m^i$ and b_m.

Convert given $\underline{x}[n]$ to new

$$x[n] = \left[\rho\left(r_1^{iT}x[n]+b_1\right)\rho\left(r_2^{iT}x[n]+b_2\right)\cdots\rho\left(r_M^{i\ T}x[n]+b_M\right)\right]^T$$

Set $\underline{w}[0] = \mathbf{0}$ and $P[0] = \mathit{l}^{-1}\,\underline{I}$ with λ = a small positive constant.

Perform the following updates for each recursive step, n.

$$P[n] = P[n-1] - \frac{P[n-1]x[n]x^T[n]P[n-1]}{1+x^T[n]P[n-1]x[n]}$$

$$g[n] = P[n]\,\underline{x}[n]$$

$$\alpha[n] = y[n] - \underline{w}^T[n-1]\,\underline{x}[n]$$

$$\underline{w}[n] = \underline{w}[n-1] + g[n]\,\alpha[n]$$

A Numerical Example

For the same data set as in the RPML case, we obtain the results shown in Figure 4.13 for the RP-RLS test case. There are 10 hidden nodes ($\lambda = 1$).

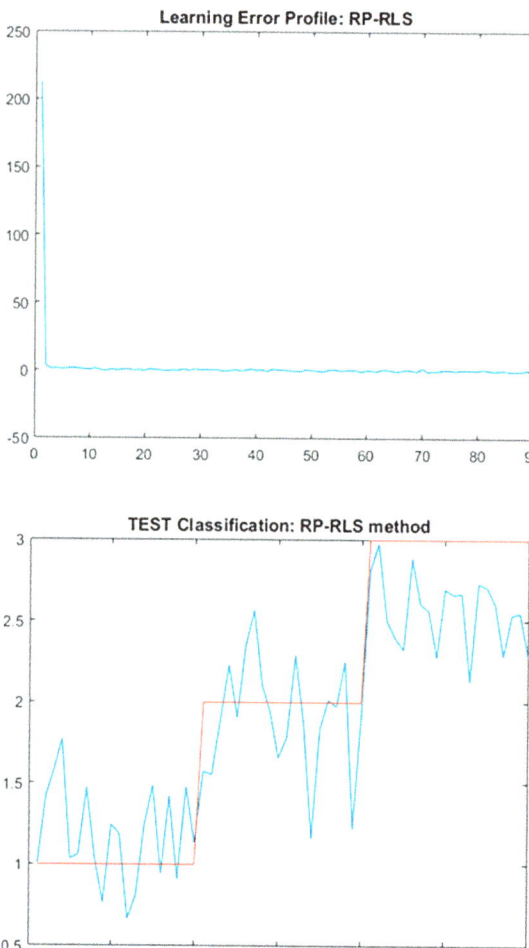

Figure 4.13 Training and test results for the RP-RLS method

Convergence performance is excellent, as is the case for the RLS. The Test MSE = 0.1631.

For each run, the random parameters of $\rho(.)$ are reset, which results in large variations in the results from run to run. This may make it difficult to choose a solution as the final one for applications.

ML ONTOLOGY

A powerful organization of concepts or the ontology of ML is based on conditional expectation. For example, let us determine the conditional expectation of Class y given the input attributes, $\underline{x}$, $E[\,y \mid \underline{x}\,]$.

Given Input $\underline{x}$, its Class, y, is the one for which $E[y \mid \underline{x}]$ is a maximum. *Implementation of the estimation of the conditional expectation above with various assumptions lead, one way or the other, to all the ML techniques.*

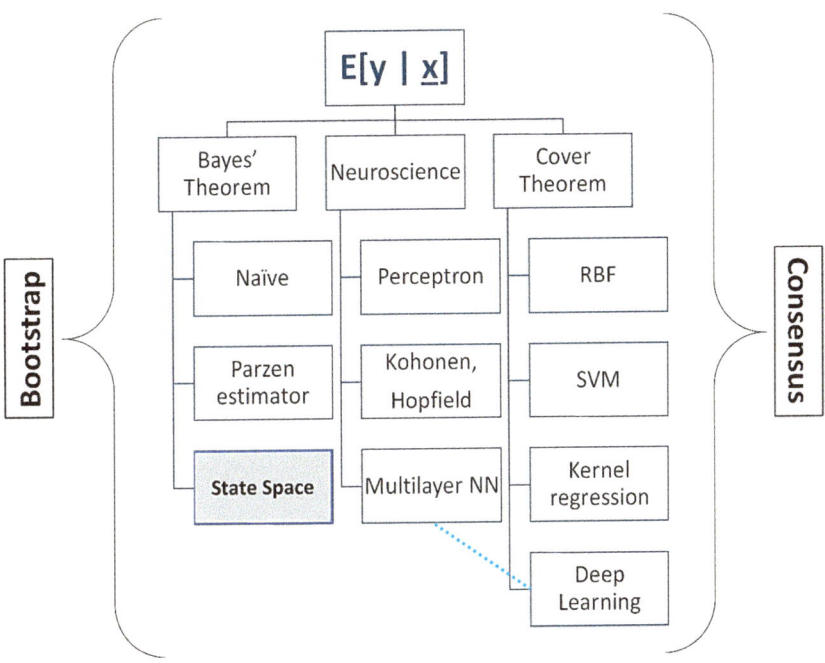

Figure 4.14 An ontology of machine learning

State-space method is the subject matter of **Part II – "Systems Analytics, the future evolution."**

CONDITIONAL EXPECTATION AND BIG DATA

In the last section, we organized ML around the conditional expectation. Here, we explain how it works with Big Data.

ML problem

Given N pairs of $\{\underline{x}, y\}$, where the values for x are the input features and y is the desired output in the Training Set, our problem is to find f where f is a non-linear function.

$$y_i = f(\underline{x}_i) + e_i \text{ where } i = 1, 2, \ldots, N \tag{1}$$

Here, e_i is an additive white noise with zero mean and variance $= \sigma_e^2$. The unknown function, f, is the conditional expectation (mean) of y given regressor, $\underline{x}$.

$$f(\underline{x}) = E[y \mid \underline{x}] = \int_{-\infty}^{\infty} y\, p_{y|x}(y \mid x)\, dy \tag{2}$$

From the joint probability density definition, we obtain

$$p_{y|x}(y \mid x) = \frac{p_{x,y}(x,y)}{p_x(x)} \tag{3}$$

The ML Solution

We have to estimate the joint density function, $p_{x,y}(x,y)$, from which the marginal density, $p_x(x)$ can be obtained by integrating out y. From the joint and marginal density, we calculate $p_{y|x}(y \mid x)$ using Equation (3). Then, the conditional expectation calculated as in Equation (1) gives us $f(\underline{x})$.

We considered the Parzen window estimation of the densities in an earlier section. In general, the density estimation is a challenging problem. In the kernel methods section of this chapter, we used kernels that had to be chosen externally and somewhat arbitrarily.

With Big Data, let us reconsider the empirical estimation of the densities required in Equations (2) and (3). *Empirical estimation is always a good first step since it includes no assumptions about the distribution and model types.* Once such a non-parametric estimate is obtained and we get to know our data better, we can bring in more sophisticated and powerful model-based estimation methods with the appropriate assumptions about the underlying probability distributions.

BIG DATA ESTIMATION

We have a very large (many millions or more) number of pairs of $\{\underline{x}, y\}$ in our Training Set. Let us use this information to empirically estimate $p_x(x), p_{x,y}(x,y), p_{x,y}(x,y)$ and $E[y \mid \underline{x}]$.

A Numerical Example

We continue using the Fisher Iris data from the earlier sections. This dataset is very small (150), but we can see the steps involved in this exercise. There are four attributes (sepal length, sepal width, petal length, and petal width, but for easy tractability, we will use only the last one.

- 1 attribute - petal width

- 1 output: 3 labelled classes - Iris: setosa, versicolor, and virginica

The Iris dataset is divided into

a. Training Set – 90 random data samples

b. Test Set – 60 random data samples

The probability density function (pdf) can be empirically estimated as a normalized histogram. Consider the one input attribute:

- Count the number of occurrences of a range of numbers in a bin.

  - Select the bin size properly. If it is too large, the histogram will consist of a few rectangles; if it is too small, there will be many zero-count bins.

- The bins versus number of occurrences in the bins form the histogram.

  - Normalize so that area under the histogram is 1, and this is an estimate of the pdf.

For the joint pdf, joint occurrences have to be counted, which can challenging. But this is just a counting exercise.

While this will be the simplest way to approach the density estimation problem, a more elegant way is to perform the multivariate kernel density (MKD) estimation. Similar to the Parzen-window estimation, the MKD code available in MATLAB performs the multivariate joint density estimation from the Training Set $\{\underline{x}, y\}$ data and the marginal density estimation from the Training Set $\{\underline{x}\}$ data.

The simplification of the Fisher Iris data problem can help our understanding of the empirical approach through the following numerical exercise.

- 1 attribute - petal width

- 1 output (same as before: 3 labelled classes - Iris: setosa, versicolor, and virginica)

As a baseline calculation, perform a simple least squares estimation of $y_i = c_0 + c_1 x_i$ using the Training Set ($i = 1$ to 90). From the estimated c_0 and c_1, we calculate the y for each of the x values in the Test Set. The results are shown in Figure 4.15 (Test Set MSE = 0.0535).

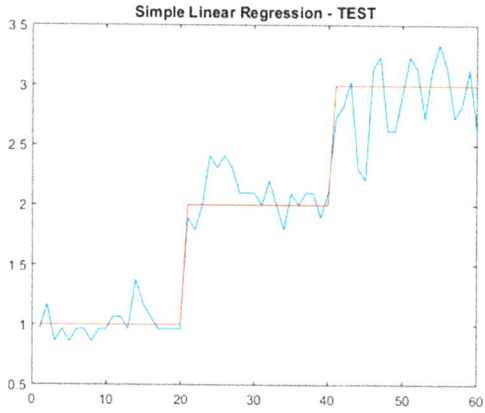

Figure 4.15 Simple least squares estimation using the Test Set

Conditional Expectation Estimation

Using the same data as above, we estimate the following:

Joint pdf – $p_{x,y}(x,y)$,

Marginal pdf – $p_x(x)$ &

$\therefore$ Conditional pdf - $p_{y|x}(y|x)$.

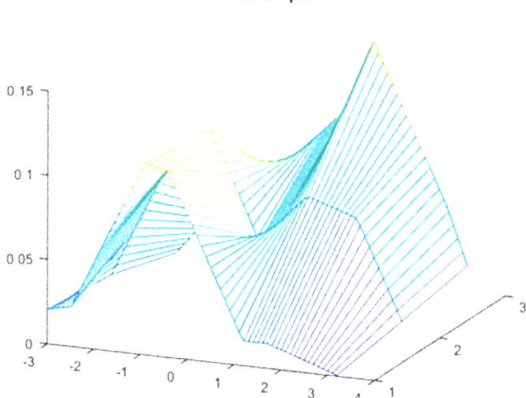

Figure 4.16 Joint probability density function from the Iris data

Training Set {x, y} is the input to the bivariate kernel density estimation function, *kde2d*, available at Matlab Central. The resulting joint pdf is shown in Figure 4.16. We have only 90 pairs of data, so the joint pdf is not as smooth as we would like but the features are still noticeable.

Training Set {x} is the input to the univariate Kernel Density estimation function, *kde*, available at MATLAB Central. The resulting marginal pdf is shown in Figure 4.17.

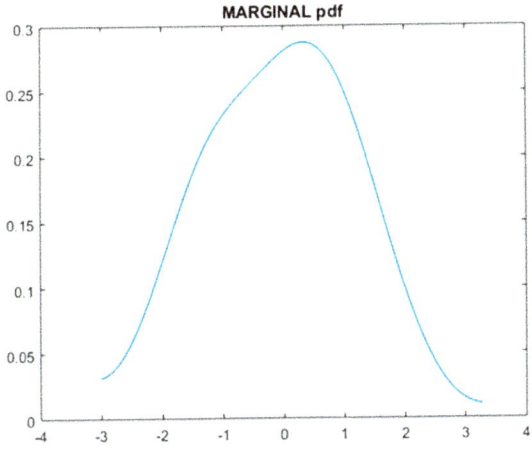

Figure 4.17 Marginal probability density function

In both cases, Multivariate Kernel Density estimation will be better than the histogram method described earlier. The discrete-valued version of the conditional expectation Equation (2) is used to obtain y (class membership) when an x input feature from the Test Set is given.

We use the following notations so that the calculation procedure is made explicit:

1. Test input – x

2. Test output – y

3. Training set input – $x^T(i)$, where i is a counter variable and $i = 1$ to 90.

4. Training set output – $y^T(i)$, where i is a counter variable and $i = 1$ to 90.

The discrete-valued conditional expectation is

$$E[y \mid x] = \sum_{i=1}^{K} y^T(i) p_{y|x}(i) \tag{4}$$

which is the counterpart of $E[y|\underline{x}] = \int_{-\infty}^{\infty} y\, p_{y|x}(y \mid x) dy$.

 Notice the similarity to Equation (5) in the Parzen window section:

$$\hat{f}(x) = \sum_{i=1}^{N} w_i\, \psi_i(x, x_i).$$

The arbitrary kernel is replaced by the empirically estimated joint pdf in Equation (4).

From the empirically-determined joint and marginal pdfs, $p_{y|x}$ is determined as follows:

For each $x(i)$ in the Test Set:

- Find each $x^T(i)$ in the Training Set equal to x. Most of the time, there will be more than one $x^T(i) = x$. Let us say there are K such instances.

- For each instance, obtain $p_{y|x}(i)$ by dividing $[x^T(i), y^T(i)]$'s joint pdf value by $x^T(i)$'s marginal pdf value.

Then, given a certain x from the Test Set, Equation (4) can be used with an additional normalization required for the expectation estimation.

$$P(i) = \sum_{i=1}^{K} p_{y|x}(i)$$

Therefore, given an $x(i)$ value from the Test Set, the classification prediction is as follows:

$$\hat{Y}\mid_x = \frac{1}{P(i)} \sum_{i=1}^{K} y^T(i) p_{y|x}(i) \tag{5}$$

These results are plotted in Figure 4.18.

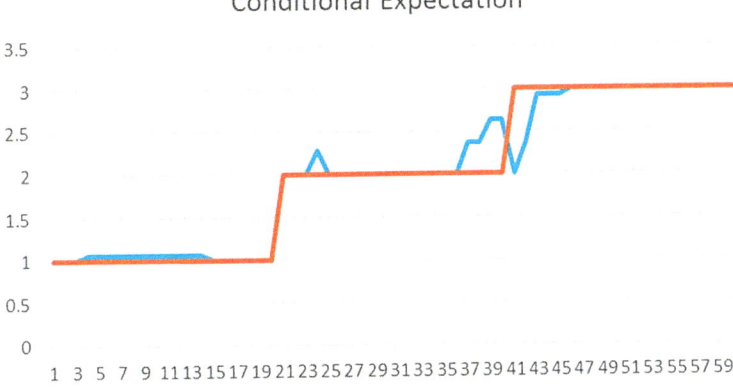

Figure 4.18 The conditional expectation results, $\hat{Y}|_x$, for each of the 60 x's in the Test Set (Test MSE = 0.0435), shown in blue

If the conditional pdf turns out to be uniform (equiprobable),

Equation (5) simplifies to $\hat{Y}|_x = \frac{1}{K}\sum_{i=1}^{K} y^T(i)$, *which is the arithmetic sample average of the Training Set y values corresponding to the Training Set* $x^T(i) = x$ *(the Test Set input). This can happen only in the unrealistic ML case of 1-to-1 mapping between x and y in the Training Set.*

MODERN MACHINE LEARNING IS BASED ON TWO KEY IDEAS:

1. Nonlinear regression and cover theorem-inspired kernel methods
2. Learning methods that arise from the Taylor series expansion of the cost or error function

---

CONCLUSION

In this chapter, we discussed a formal framework for learning and how we can assemble various ML topics into a coherent whole.

A significant amount of information was covered in this chapter, so it is recommended that you refer back to the previous chapters and write the steps for the solutions to problems, which is a much better way to develop your understanding than looking through the MATLAB m code. Once you have written out the steps, the MATLAB code can be helpful in understanding further details.

This MATLAB code is found on the companion files or by writing to the publisher at *info@merclearning.com.*

ADAPTIVE MACHINE LEARNING

In Part II, we focus exclusively on incorporating dynamics into machine learning. Descriptive analytics gives a snapshot of the history of your business. It is very useful in getting a data-driven picture of a part of your business. But, as we saw in the first chapter, for business to perform at a high level requires them to be close to the "edge of complexity overload." It is predictive analytics that helps businesses be competitive using data.

Prediction is challenging, mainly because systems are not stationary. For random processes, non-stationarity is a well-defined property. In common parlance, it means that things change over time, sometimes gradually and sometimes abruptly. Hence, the need for constant tracking.

As mentioned earlier, business solutions require ongoing attention. Predictive analytics solutions should be adjusted and re-applied on a regular basis. Then we can track the system and continually update the predictions to keep false positives and negatives in check. This explains the need for dynamics in machine learning.

WHAT IS DYNAMICS?

Statics and dynamics are two topics in physics. There is something similar in most fields; for example, there are context-free and context-sensitive grammars (the location of a word in a sentence is sensitive to the meaning as well as being grammatically correct). Taking the generalized meaning of dynamics (and separating it from meaning "time variation" only), we consider dynamics as referring to the property of any quantity (data or features) that are ordered over one or more independent variables. It is notable that such an ordering comes with correlations among data, which can be exploited for better results (ordered data can also be uncorrelated to each other; then each new datum is an "innovation" or purely new information).

Typically, the independent variable is time. However, the independent variable can be space (in two- or three-dimensions). It can also be the position in a sentence where the word location has significance (if you switch the subject and predicate order, the English grammar parser will fail). In business cases, repeated data measurement for tracking clearly constitutes dynamics.

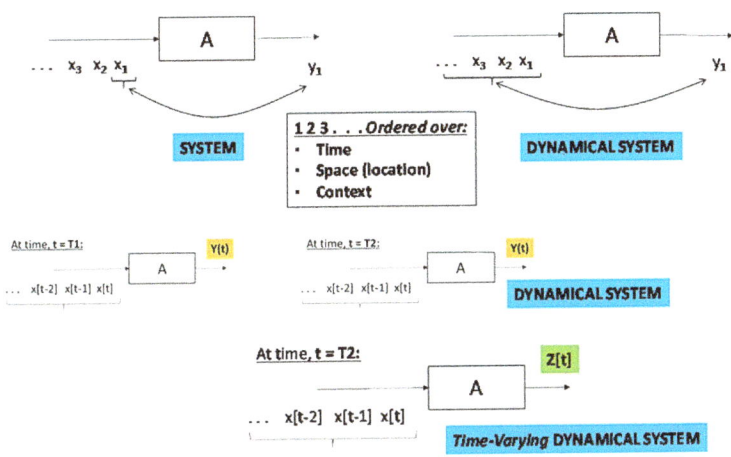

Figure 4.19 Dynamical system explanation

In the systems context, a dynamical system is the object of study. We take a brief detour to remind ourselves of what dynamical and time-varying dynamical systems mean (the x and y values in the diagram can be scalar or vector).

On the top-left of Figure 4.19, we have the familiar static system in the sense that input and output at the present time are related. This is the relationship captured in a linear or multiple-linear regression model (when the coefficients of the regression equation are constants and there only x_1 values that are scalar or vector). This approach is the subject matter of basic statistics. (See Chapter 2 for various forms of linear and non-linear regression equations.)

In the systems approach, we note the following:

- If the input ordering (along any independent dimension) is relevant, it is a dynamical system.

- On the top-right of Figure 4.19, this implies that System A has "memory" or "energy storage" elements in it.

- As a practical matter, this will mean that the output y_1 varies along the independent dimension (hence, the term "dynamical").

- Time-varying Dynamical Systems

  - Consider "time" as the ordering principle.

  - The middle row in Figure 4.19 shows the dynamical system on the top-right now applied at time $= T1$ and $T2$. The output is unchanged $= Y(t)$.

  - If the outputs at $T1$ and $T2$ are different as shown at the bottom (at $T2$, output is $Z(t)$ and not $Y(t)$), the system is time-varying.

  - This means that parameters of A are changing over time. In the random process case, this is called *non-stationary*.

- If the parameters of A are non-linear, it is a non-linear dynamical system. Remember that in the regression equation case (Chapter 2), the coefficients have to have x's with non-linear relationships with y.

In Part II, our focus is on dynamical and time-varying dynamical systems. The basic mathematical framework to study dynamics is differential equations (or difference equations, for discrete-time systems). In Part II of this book, we cast dynamical machine learning within the framework of systems theory, which is built upon calculus, to study, predict, and control both natural and man-made systems. We call such an approach to ML *adaptive machine learning*.

REFERENCES

[HS08] Haykin, S, *Neural Networks and Learning Machines*, Pearson, 2008.

[AY12] Abu-Mostafa, Y, *Learning from Data*, AML Books, 2012.

[HG12] Huang, G, et al., *Extreme Learning Machine for Regression and Multiclass Classification*, IEEE Transactions on Systems, Man, and Cybernetics, 2012.

SYSTEMS ANALYTICS

CHAPTER 5

SYSTEMS THEORY FOUNDATIONS OF MACHINE LEARNING

Machine learning today tends to be "open-loop," which means we need to collect a significant amount of data offline, process the data in batches, and generate insights for eventual action. There is an emerging category of ML business use cases that are called "In-Stream Analytics (ISA)."

INTRODUCTION-IN-STREAM ANALYTICS

Here, the data are processed as soon as they arrive and insights are generated quickly. However, action may be taken offline and the effects of the actions are not immediately incorporated back into the learning process; this is an example of a "closed-loop" system, and we call this approach *systems analytics* or *adaptive machine learning* (AML).

Why is Time Important?

Here is a small list of business use cases summarized where "time" is important.

1. Fraud Detection
   - Rules and scoring based on historic customer transaction information, profiles, and even technical information to detect and stop a fraudulent payment transaction

2. Financial Markets Trading
   - Automated high-frequency trading systems

3. IoT and Capital Equipment Intensive Industries

- Optimization of heavy manufacturing equipment maintenance, power grids, and traffic control systems

4. Health and Life Sciences

- Predictive models monitoring vital signs to send an alert to the right set of doctors and nurses to take an action

5. Marketing Effectiveness

- Detect mobile phone usage patterns to trigger individualized offers

6. Retail Optimization

- In-store shopping patterns and cross selling; in-store price checking; and creating new sales from product returns

One factor that necessitates real-time interaction is the closed-loop nature of these use cases. In every case, an external event happens. An analytics module determines a recommended action and creates a response that impacts the external event in a timely manner.

What "real-time" means is case dependent. The rate at which data collection, analysis, and action happen could be milliseconds, hours, or days. The industry name for this is *Event Stream Processing* or *In-Stream Analytics* (ISA). There are specific implementations of computer systems, databases, and data flow protocols available to address the stringent requirements of ISA systems.

It is desirable that when the analytics module determines a recommended action and creates a response that will impact the external event, you also measure the i of the action so that

1. we know if our action was good,

2. we can use any shortcomings to improve the ISA so that the next ISA event will have a better outcome and

3. we can attribute the right portion of the result to ISA's action.

EXPLICIT DEFINITION OF LEARNING

If learning is the process of the "generalization from experience," we can be more explicit and say that learning is the generalization of past experience and the results of new action.

With this broader view of learning, all analytics applications conform to the picture above, other than purely descriptive analytics. ML algorithms generate outputs that requires action. The learning system may be kept hidden from users, but any ML application that has a business impact will have to "close the loop." Off-line ML systems (that operate without being seen by users) still require a "trickle" of continuous learning lest the learned system becomes old and unresponsive to new changes. Even a language translation ML system belongs in the closed-loop category for proper continuous operation (for example, new words and expressions enter the lexicon all the time).

In summary, adaptive ML in a closed-loop configuration is a basic feature for any ML business solution.

BASICS FOR ADAPTIVE ML

In the first part of this book, we approached machine learning as follows:

1. Given N pairs of $\{x, y\}$, where x is the input feature and y is the desired output in the Training Set, find the function, f:

$y_i = f(x_i) + e_i$ where $I = 1, 2, \ldots, N$ where e_i is an additive white noise.

The function, f, that maps x_i to y_i can be

- Set up as a regression problem and using least squares, one can find the function as the best-fit curve in some mean squared sense.

- For the case of functions that are linear, the recursive linear regression algorithm provides the exact solution in a step-by-step manner as a new pair of $\{x, y\}$ arrives.

- We also saw in the non-linear case how this problem can be set up as a kernel regression problem:

$$\hat{f}(x) = \sum_{i=1}^{N} w_i \psi_i(x, x_i) \text{ for } i = 1, 2, \ldots, N.$$

- Here we combine various non-linear shapes (kernels) to find the best hyper-surface that maps x_i to y_i.

2. Another approach we took was the consider the mapping problem as a conditional expectation problem:

$$f(\underline{x}) = E[y \mid \underline{x}]$$

- From the Training Set of $\{x, y\}$ pairs, we estimate the joint density from which we get the necessary marginal density. We then get the condition density, which allows us to compute the conditional expectation.

$$E[y|\underline{x}] = \int_{-\infty}^{\infty} y\, p_{y|x}(y \mid x)\, dy$$

- In the ML ontology section (shown again in Figure 5.1), we found that there are many methods to estimate the conditional expectation, including state space method, which we discuss in detail in this chapter.

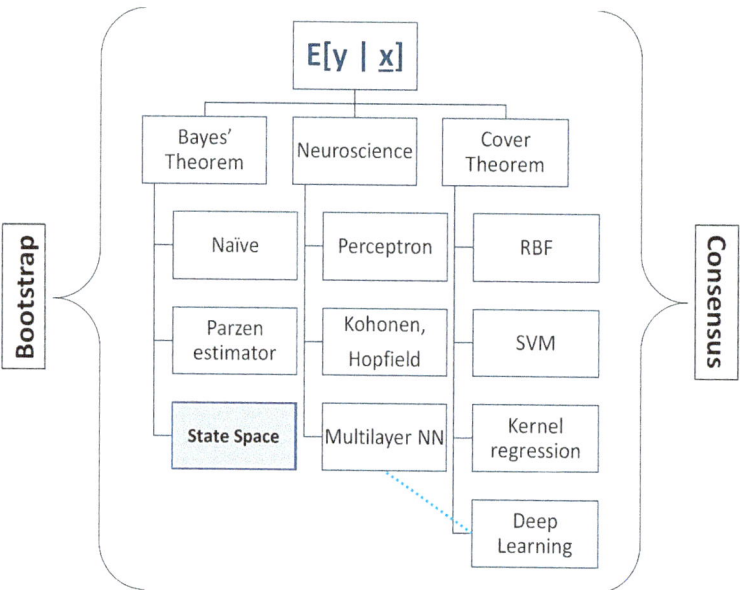

Figure 5.1 Machine learning ontology

We develop the adaptive ML solutions in a principled manner in this chapter. While we work through the necessary background for the state space theory and Bayes estimation, it should be noted that these topics are rich in history, deep in theory, and wide in applications. An independent study of these topics will strengthen your grasp of the topics discussed in Part II.

EXACT RECURSIVE ALGORITHMS

Let's return to the earlier discussion about off-line and closed-loop processing. Off-line methods can use batch processing but closed-loop, ISA will require recursive online processing so that each data input is processed as it arrives. The nature of recursive algorithms can be confusing and is often misunderstood in the context of machine learning.

In Chapter 4, we encountered the recursive least squares (RLS). At that time, we did not elaborate the exact nature of RLS; it is exact in the sense that the batch least squares solution using all the data at one time and the solution obtained by RLS are identical.

Let us consider the case of finding the mean height of students in a class. We decide to use the standard formula for the sample mean. We measure the heights of N students and apply the following estimate:

$$\overline{x}_N = \frac{1}{N} \sum_{i=1}^{N} x_i$$

where x_i is the height of an individual student. With simple algebraic manipulations, we can write the sample mean estimate as a recursive algorithm so that the sample mean is calculated as soon as each individual student's height is measured.

$$\overline{x}[n] = \frac{x[n]}{n} + \frac{(n-1)}{n} \overline{x}[n-1] \text{ with } \overline{x}[n] = 0 .$$

Here n is a counter variable. When $n = N$, the usual sample mean and the recursive sample mean are exactly equal. It is in this sense that recursive algorithms are exact. We can re-write the exact recursive algorithm (ERA) for the sample mean to make it easy to compare and contrast with learning algorithms in Part I (especially the discussion in Chapter 4):

$$\overline{x}[n] = \frac{(n-1)}{n} \overline{x}[n-1] + \frac{x[n]}{n}$$

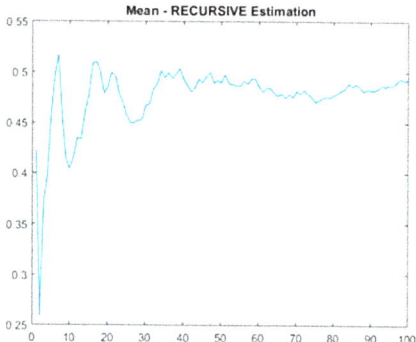

Figure 5.2 Recursive estimation of mean value

Here, we can think of the first term on RHS to be the past estimate of the sample mean and the second term, a correction term. Compare this example to the steepest descent learning formula in Chapter 4. The learning curve is given in Figure 5.2.

$$\Delta w[n] = -\eta\, \frac{\partial E[n]}{\partial w[n]} = +\eta e[n]\, \frac{\partial \hat{y}[n]}{\partial w[n]},$$

where η is the "step-size".

This can be written in terms of updated weights:

$$w[n+1] = w[n] + \Delta\, w[n] = w[n] + \eta e[n]\, \frac{\partial \hat{y}[n]}{\partial w[n]}.$$

This formula has the same form as our sample mean ERA above: it is the past estimate plus a correction term. There is a difference, however. ERA is in the form of a filter, where the correction term is based on the current value and updates the current sample mean. However, in the steepest descent formula, the correction term is based on the current value and updates the future weight (the prediction form).

The RLS is an exact recursive algorithm (ERA), which applies when the least squares problem is linear-admitting exact solution at each step. In the non-linear case, we resort to steepest descent where $f(.)$ is non-linear in the model, $y_i = f(x_i) + e_i$. Then, the error surface is not smooth (quadratic) and an approximate learning method is needed, where it searches the error surface to find local or global optima.

The ERA is a learning method. Solving the whole problem up to a certain step is the same as learning "as much as it can" from each new piece of information, keeping in mind the linear constraint. It is well-known in systems theory and signal processing that many methods can be cast as an ERA because many problems are solutions of the Normal equation mentioned in Chapter 4. This equation is as follows: $\underline{w}* = (\underline{x}^T\underline{x})^{-1}\underline{x}^T \ \underline{y} = \underline{R}^{-1}\underline{r}$, where $\underline{R}$ is the autocorrelation matrix of input, $\underline{x}$, and $\underline{r}$ is the cross-correlation vector between input, $\underline{x}$, and desired output, $\underline{y}$. When the $\underline{R}$ matrix inversion is replaced by the matrix inversion lemma, the result is an exact recursive algorithm, which solves the Normal equation exactly at each step with the information available until that step.

From the sample mean ERA discussion above, you would have noticed that storage requirements are drastically less for ERA since only the past average and current data is to be processed. For a very large value of N, this could be a significant implementation factor. Memory savings is the least of the desirable properties for ERA.

EXACT RECURSIVE ALGORITHMS (ERAS)

ERAs solve the estimation problem as each new piece of data arrives. This ability to track data is the value of ERA algorithms. In other words, if $f(.)$ in our model, $y_i = f(x_i) + e_i$, is time-varying, ERAs have the ability to track the changes because the exact solution is calculated at each step. In practice, the nature of variation (speed of change and complexity) will limit the ability of ERAs be fully change-adaptive. The recursive least squares (RLS) is an example of exact solutions for linear time-varying least squares models.

Machine learning is not focused on tracking solutions. Why is that? As we saw in the last section, the business use cases that demand closed-loop solutions may be assumed to be non-varying in time or space or other dimensions as a simplification but in reality, every model evolves slowly or quickly. Any serious solution must be adjusted periodically and performed again on a regular basis. The entities involved vary (over time, in this case) and tracking will improve the results. It may even be necessary on an ongoing basis if the changes are significant.

ERAs adapting to change can be categorized as leading to a model that changes to a "new normal" or changes to "abnormal." From a steady-state, an ML algorithm learns what represents normal (think of an IoT example where a piece of machinery is operating correctly). When deviations occur, ERAs quickly learn the new model. There are two ways to flag this change:

a. Monitor the ERA model parameters for change.

b. Monitor the output (such as prediction error) of a non-tracking algorithm; the change in the system causes a larger-than-normal error in the monitored output.

As you can see, both options flag an event but the ERA is more informative. It characterizes the change (in terms of new model parameters) which is valuable to know. It tells us whether the system has changed to a new normal or changed to abnormal.

The response of the human operator to either option is different. The accumulated information about the new normal adds to the knowledge base, which allows us to develop more advanced versions of the ML solution. This learning at a higher level enhances the power of ML solutions as time goes on. We call this an adaptive ML solution. Thus, ERAs lead to adaptive ML. Once we incorporate non-linear models into our system, we have a complete solution to AML problems.

IN PART II, WE FOCUS ON AML (ADAPTIVE MACHINE LEARNING)

Its features are:

1. Tracking or time-varying (or over any other dimension such as space)l

2. Non-linear

3. Exact recursive algorithms

AML methods are necessary for closed-loop solutions. Off-line or batch methods are suitable for investigations and explorations, but the solution to any business problem will require it to be closed-loop with the time available between event and action varying from milliseconds to hours, days, or months. The adaptive nature of AML, where the solution adapts to the changing environment, renders the solution fully automatic, removing the need for human intervention.

A Practical Example of AML Using ERA

To consolidate the concept of adaptive ML, let us consider a practical (but fictional) ML problem related to the Fisher Iris data.

Case Study of the Fisher Iris Farm and Automated Flower-Sorting Systems

Figure 5.3 shows a simple version of the Fisher Iris Farm's industrialized flower sorting system. Big heaps of three types of iris flowers are dropped off by the farm equipment on the left side of the system. The automatic sorting system has to classify each flower as belonging to setosa, virginica, or versicolor. It then sends them on to three separate conveyer belts so that the types can be packaged separately.

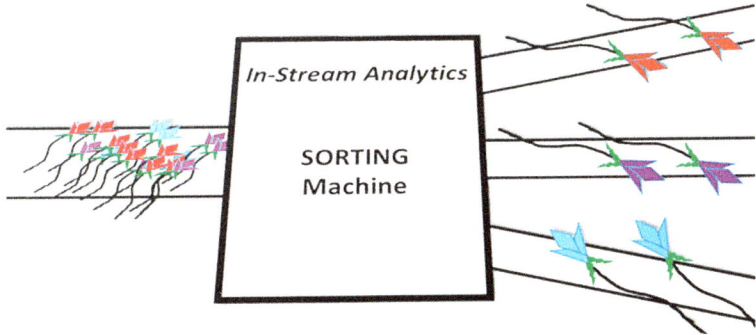

Figure 5.3 On-line flower sorting machine

We have an optical system that can measure three attributes (sepal length, sepal width, and petal length) automatically while the flowers pass through the sorting machine.

Step 1: Offline classifier development

- We collect 150 sets of {Attributes, Iris Type}.

- Using 90 pairs of data, we train a classifier using the RLS as we did when we solved the Iris problem in Chapter 4.

- We test the classifier using 60 "held-out" pairs and find that the classification accuracy is acceptable for this industrial application.

Step 2: Online operation

- The optical system in the sorting machine automatically measures the sepal length, sepal width, and petal length of the passing flower.

- This data in passed on to the ISA system, which uses the classifier trained in Step 1 above.

- An arm mechanism directs the right flower onto the right conveyer belt for that flower.

As we know, the classifiers are not perfect and there will be misclassifications occasionally. For business reasons (the wrong flower in the wrong box generating customer complaints), the data scientist is asked to improve the performance.

Step 3: Performance improvement using ERA

- The quality control department inspects the operations for 30 minutes during the day at random times.

- When a flower is on the wrong conveyer belt, a switch is pressed that informs the ISA automatically of the wrong classification, and the ISA extracts the attributes for that flower from the logs.

- This information is then used by the exact recursive algorithm to update the classifier online.

This system operates as follows:

Offline Training → Operation → Closed-loop feedback → Recursive Online Update of the Classifier.

If there was a continuous automatic quality control system, the closed-loop feedback and recursive online update can be done continuously with the classifier improving each time. In addition, suppose that attributes evolved due to a dry spring or because of a new fertilizer, the recursive online update will track these changes, learn online, and maintain the performance of the classifier.

Clearly, there are other applications of ML where insights are generated and resulting report is submitted to the business executive. Here, the loop is not closed on-line, but as ML use by the business matures, the executive will want the insights to be applied to the business problem. The executive will also want to know if the action produced good results and if so, what portion of the good result can be attributed to the ISA.

Therefore, the need to close the loop is unavoidable for any sustainable ML business application. The conceptual architecture for a practical adaptive machine learning system is shown in Figure 5.4.

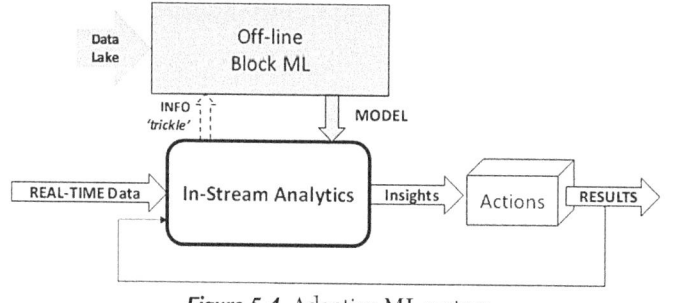

Figure 5.4 Adaptive ML system

CHAPTER 6

STATE SPACE MODEL AND BAYES FILTER

I n this chapter, we discuss the state space model and Bayes filter. We begin with the state-space basics. The function relationship, $y_i = f(x_i) + e_i$, can be elaborated upon in the following manner. In the linear case, an alternate way of writing the relationship in vector form is $y = w^T x + e$, with compatible dimensions. If $\{x, y\}$ were made available to us one at a time, we can add a discrete-time index, n:

$$y[n] = w^T[n]\, x[n] + e[n] \tag{1}$$

This conforms to our familiar LTI system diagram with noise added.

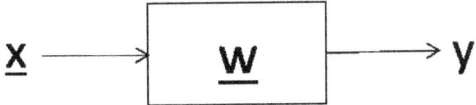

Equation (1) can be re-written as follows:

$$w[n + 1] = w[n]$$
$$y[n] = w^T[n]\, x[n] + e[n] \tag{2}$$

These equations are identical to Equation (1). The only reason to use this form is that Equation (2) conforms to the standard state-space model equations. The general form of state equations is

$$s[n + 1] = A\, s[n] + C\, x[n] + \mathbf{v}[n]$$
$$y[n] = B\, s[n] + e[n] \tag{3}$$

When we compare Equations (3) and (2), s, the state $= w$, the regression weights; x is the input; the state transition matrix then is $A = I$; and the dynamic noise $v = 0$. $B = x[n]$ is the measurement matrix, C (= 0) is the input matrix, and e is the Measurement noise. The A and B matrices can be time-varying. Equation (3)'s notation can be changed so that the state-transition and measurement functions are non-linear. The state, s, will have the Markovian property such that the current state is only dependent on the immediate past state. States are also hidden in that only their effects can be seen through the output, y.

Why are we using the more cumbersome notation in Equation (3)? We use this approach so that we can reuse the well-developed linear or non-linear and Gaussian or non-Gaussian time-varying solutions to our problem of estimating the conditional expectation of the class, y, given input, x. Here, it becomes the problem of obtaining the conditional expectation of the state, $\underline{s}$, given $\underline{y}$ & $\underline{x}$.

Therefore, if we are able to evaluate $E[\underline{s}[n + 1] \mid \underline{y}[n], \underline{x}[n]]$, we can use Equation (3) when a new input, $\underline{x}[n + 1]$, arrives to calculate the output, $\underline{y}[n + 1]$, since the updated state is already available from the previous step.

In dynamical systems, a quick level-set is $a_1\ddot{x} + a_2\dot{x} + a_3 x = 0$, where the values for a_i are constants in a linear differential equation in x. Then, we have $b_1\dot{x} + b_2 x^2 = 0$, where the values for b_i are constants is a non-linear differential equation in x. Whereas

$$a_1(t)\ddot{x} + a_2\,\dot{x} + a_3(t)x = 0$$

where a_1, a_3 are varying over t (and a_2 is a constant), and this equation is a time-varying differential equation in x. We also know the following are true:

- We need not specify t here as a time variable, but just as a "dimension" over which the coefficients of the differential equation are varying.

- Similarly, we need not specify "time" as the derivative with respect to which $\ddot{x}$ and $\dot{x}$ are defined.

- This flexibility allows us to proceed if t is "space," or if t is the word-order such that the equation captures the context-sensitivity of a word in a sentence.

Unfortunately, there are no common names for such equations. Often, t is implicitly assumed to be "time" and we call such equations *time-varying differential equations*. There is a conjecture that any non-linear differential equation can be equivalently replaced by a time-varying linear differential equation [YP11]. Past experience indicates that this is true in the case of time-varying spectrum estimation or time frequency distributions [MP96].

Though linear, time-varying coefficients bring some issues to the solution of such differential equations, yet they are far easier to handle than non-linear differential equations. This is our express reason to accept the conjecture. We define t (or its discrete-time counterpart, n) as a *counter variable*.

With this discussion in mind, we can re-write the state equations in Equation (3) as .

$$s[n + 1] = A[n]\,s[n] + C[n]\,x[n] + v[n]$$

$$y[n] = B[n]\,s[n] + e[n] \tag{4}$$

This is our discrete *Time-Varying Linear* (TVL) state-space model.

The flow graph of our TVL state-space model is shown in Figure 6.1.

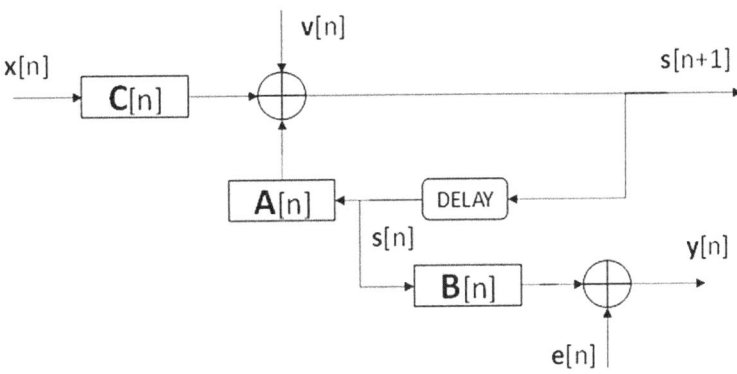

Figure 6.1 Flowchart of state-space model

A Numerical Example

Early in Chapter 2, we considered a multiple linear regression model for the Fisher Iris problem where we used three features to classify the following:

Class = 2 + 1.9953 *Feature1 – 0.026357*Feature2 –
0.4436*Feature3 + noise

Class = Class – 2 = 1.9953 *Feature1 – 0.026357*Feature2 – 0.4436*
Feature3 + noise

If we write these statements in general terms, we obtain

$$y[n] = a_1 x_1[n] + a_2 x_2[2] + a_3 x_3[n] + e[n]$$

where $a_1 = 1.9953$; $a_2 = -0.026357$; $a_3 = -0.4436$

If we write the same equation as our TVL model in Equation (4), we obtain

$$B^T[n] = \begin{pmatrix} x_1[n] \\ x_2[n] \\ x_3[n] \end{pmatrix}, \quad \underline{A}[n] = \underline{I}, \ \underline{C}[n] = \underline{0}$$

The TVL model equations for this problem are as follows:

$$s[n+1] = A[n]\, s[n] + v[n]$$

$$y[n] = B[n]\, s[n] + e[n]$$

This equation has a simplified form compared to Equation (4). The $\underline{B}[n]$ matrix is the input vector, $\underline{x}[n]$. Over n,

$$\underline{s}[n] = \begin{pmatrix} a_1[n] \\ a_2[n] \\ a_3[n] \end{pmatrix}$$ and state, $\underline{s}$, will converge to $[a_1 = 1.9953\ a_2 = -0.026357$

$a_3 = -0.4436]^T$.

STATE-SPACE MODEL OF DYNAMICAL SYSTEMS

After the preparatory discussions in the last few sections, let us formally assemble the various concepts and define our state-space model, which leads to a Bayesian estimation of the entities of the model. We perform the development side-by-side with regression models, which is just one of the many data models.

Multiple Linear Regression Model:

Model: $y[n] = a_0 + a_1 x_1[n] + a_2 x_2[n] + \ldots + a_M x_M[n] + w[n]$

$\{a_i\}$ are unknown model parameters and w is white noise with a zero-mean and specified variance.

Given: $\{y, x\}$, Training Set

Find: For a new given x^{NEW}, find y^{NEW}.

Solution: Our solution uses the least squares approach. We use the recursive theme, and so utilized the recursive least squares approach on the Training Set to find the model parameters. Then, we substitute x^{NEW} into the model equation to get y^{NEW}.

Algorithm: Use the RLS algorithm from Chapter 4.

Given: $\{\underline{x}[n],\ \underline{y}[n]\}$ for $n = 1$ to N and with a$[0] = 0$ and $\underline{P}[0] = \lambda^{-1}\underline{I}$ with $\lambda =$ a small positive constant, perform the following updates for each recursive step, n.

$$\underline{P}[n] = \underline{P}[n-1] - \frac{P[n-1]x[n]x^T[n]P[n-1]}{1+x^T[n]P[n-1]x[n]}$$

$$g[n] = \underline{P}[n]\,x[n]$$

$$\alpha[n] = y[n] - a^T[n-1]\,x[n]$$

$$\underline{a}[n] = \underline{a}[n-1] + g[n]\,\alpha[n]$$

State-Space Model

Model: We have seen a few examples in the last section, but we will choose a minimal model focused on the machine learning problem. Again, keep in mind that the following is a different data model independent from the multiple linear regression.

$$s[n] = A[n]\,s[n-1] + B[n]x[n] + v[n]$$

$$y[n] = H[n]\,s[n] + e[n] \tag{5}$$

Here, $s[n]$ is the state at counter $=n$ and $A[n]$ is a known, varying matrix (in some cases, $= I$, identity matrix). The input matrix, $B[n]$, is a known matrix and the observation matrix, $H[n]$ is another known matrix. v and e are the white noise with a zero-mean and specified variance. (Though multiple linear regression and state-space models are two independent models, one can see that if $B[n] = 0$ and $H[n] = (1\ x_1[n]\ x_2[n]\ \ldots\ x_M[n])$, the multiple linear regression model parameters, $a = s$, the states.)

Given: $\{y, x\}$, Training Set

Find: For a new given x^{NEW}, find y^{NEW}.

Solution #1: Conditional expectation. Find $p_{y|x}(y\,|\,x)$; from which, find $E[y^{\text{NEW}}\,|\,x^{\text{NEW}}]$.

Solution #2: The algorithms first estimate the states, s, i.e., $p(s|y,x)$ and then $E[s|y]$. These conditional expectations of states are substituted in Equation (5) and for x^{NEW}, we estimate y^{NEW} as $E[y^{\text{NEW}}|x^{\text{NEW}}]$.

Algorithm: We will identify the algorithm for Solution #2 after discussing different types of Bayes estimates.

Bayes, MAP, and Maximum Likelihood Estimates

Let us briefly consider and explicitly tie the three estimates in this section title together. From Bayes theorem discussion, we know the following.

From joint pdf definition, $p_{y\,x}\ y\mid x\quad\dfrac{p_{x\,y}(x,y)}{p\ x}$

We can replace joint pdf and obtain the Bayes Theorem:

$$p_{y|x}(y\mid x)=\frac{p_{x|y}(x\mid y)}{p_x(x)}\,p_Y(y)$$

$$p(y\mid x)=\frac{p(x\mid y)}{p(x)}\,p(y) \tag{a}$$

We simplify the notation by eliminating the subscripts for ease of presentation.

Maximum Likelihood (ML) Estimate

$p(x)$ is a normalizing term that is ignored. $p(y)$ is assumed to be uniformly distributed.

$$\underbrace{\arg\max}_{Y} = p(x\mid y) \tag{b}$$

Equation (a) yields the value of y for which the function is a maximum for a given x.

Naïve Bayes

With the same assumptions as made for the ML estimation plus the conditional independence assumption, we have the following:

$$p(x\mid y)\cdot p(y)=p(x,y)=p(x)\cdot p(y)$$

$$\underbrace{\arg\max}_{Y}=p(x)\cdot p(y)$$

Maximum A-Posteriori (MAP) Estimate

$$\underbrace{\arg\max}_{Y}=p\big(y\mid x\big) \tag{c}$$

Equation (b) yields the value of y for which the function is a maximum for a given x. The normalizing term, $p(x)$, acts as a regularizer for the MAP estimate. However, if there is no a-priori information and normalizing term is ignored, Equation (c) gives the same result as the ML estimate in Equation (b).

Bayes Estimate or Conditional Expectation

$$\mathrm{E}[y|\underline{x}] = \int_{-\infty}^{\infty} y\, p(y\,|x)dy \qquad\qquad (d)$$

In Equation (d), instead of a point estimate (the maximum value of the conditional pdfs), the conditional expectation is a moment estimate or an average value calculation. This is the Minimum Mean Square Error (MMSE) estimate; this estimate is robust.

KALMAN FILTER FOR THE STATE-SPACE MODEL

There are two approaches to developing a recursive algorithm to estimate the state-space Equation (5) in the last section: the Bayesian approach and innovations approach. The former is probabilistic [SS13] and the latter is information-theoretic [KT00]. Both approaches to the derivation result in the same algorithm: the Kalman filter. Derivation via the Bayesian approach requires the assumption that e, $v = N(0, R)$; i.e., e and v are distributed as Normal (or Gaussian) with mean = 0 and covariance = R. The innovations approach is not probabilistic and hence does not require a Gaussian assumption.

Let us re-state Equation (5):
$$s[n] = A[n]\, s[n-1] + B[n] \times [n] + v[n]$$
$$y[n] = H[n]\, s[n] + e[n] \qquad\qquad (5)$$

The Kalman filter is the closed-form, exact solution to the state-space equations:

- n – Counter that goes from 0, 1, 2, . . . , $n-1$, n
- s – State, $\in R^K = N(s[0], P[0])$ where $s[0]$ and $P[0]$ are the initializing mean and covariance
- x – Input, R^N
- A – Transition matrix
- B – Input matrix
- v – Process noise = $N(0, R_V)$
- y – Measurement, $\in R^M$
- H – Measurement matrix
- e – Measurement noise = $N(0, R_E)$

- *The evolution of the state is Markovian in that $p(s[n]|$ $s[n-1], s[n-2], \ldots, 2, 1, 0) = p(s[n]|s[n-1])$.*

- *The process and measurement noise are statistically independent.*

- *$N(m,R)$ – Normal or Gaussian distributed random vector with a mean $= m$ and covariance $= R$.*

We want to find $p(s[n]|y[n])$, which we note is the same as $p(s[n]|y[n], x[n])$. However, $x[n]$ disappears from the rest of the probability calculations since they are deterministic, i.e., $\{x\}$ in the Training Set and x^{NEW} are always given beforehand.

$$p(s[n]\,|\,y[n]) = \frac{p(y[n]\,|\,s[n])}{p(y[n])} p(s[n])$$

Using Markovian property and Gaussian assumptions in the estimation of $p(s[n]|y[n])$, it can be shown that Bayesian Filtering equations are as follows.

$$p(s[n]|y[n-1]) = N(s[n]\,|s_M[n|n-1], P[n|n-1])$$

Gaussian with PREDICTED mean $s_M[n|n-1]$ and PREDICTED covariance $P[n|n-1]$.

$$p(s[n]|y[n]) = N(s[n]\,|\,s_M[n|n], P[n|n]),\ \text{FILTERED mean } s_M[n|n] \text{ and}$$
covariance $P[n|n]$.

$$p[y[n]|y[n-1]] = N(y[n]\,|\,H[n]s_M[n|n-1], S[n]).$$

The parameters of the Bayesian Filter above ($s_M[n|n-1]$, $P[n|n-1]$, $s_M[n|n]$, $P[n|n]$ & $S[n]$) can be calculated by Kalman filter prediction. We can therefore update the steps as follows:

Initialization: $s_M[0|0] =$ small random values; $P[0|0] = \lambda^{-1}I$ with $\lambda =$ small positive constant

At Counter = [n]: *State Prediction: From State at $[n-1]$*

$s_M[n|n-1] = A[n]\, s_M[n-1|n-1] + B[n]\, x[n]$: State Estimate Prediction

$P[n|n-1] = A[n]\, P[n-1|n-1]A^T[n] + R_V[n]$: Covariance Estimate Prediction

State Update: Using Measurement at $[n]$

$q[n] = y[n] - H[n]\, s_M[n|n-1]$: Innovation

$S[n] = H[n]\, P[n|n-1]H^T[n] + R_E[n]$: Innovation covariance

$$K[n] = P[n|n-1]\, H^T[n] S^{-1}[n] \qquad\quad : \text{Kalman gain}$$

$$P[n|n] = P[n|n-1] - K[n]S[n]K^T[n] \qquad : \text{Covariance estimate a posteriori}$$

$$s_M[n|n] = s_M[n|n-1] + K[n]q[n] \qquad\quad : \text{State estimate a posteriori}$$

At convergence, $s_M[n|n]$ = conditional expectation, $E[s|\,y, x]$.

- Use $s_M[n|n]$ as $s[n]$ in the state-space Equation (5) with noise terms, v, $e = 0$.

- For x^{NEW}, compute y^{NEW}, which is the Bayesian estimate.

An example application of the Kalman filter for dynamical machine learning is as follows.

In Chapter 4, we explored an empirical approach to estimate $E[y|x]$ using Big Data. With the availability of large quantities of Big Data (which are *homogenous*), this empirical approach, which is free of distribution (Gaussian) and linearity assumptions, is the best the first step. In our dynamical ML implementation framework shown in Figure 5.4, the empirical estimation method can be exploited for an off-line, block ML solution.

For the in-stream portion, however, we need a recursive solution that updates the off-line solution as each new datum arrives without having to reprocess all the historical data. In the recursive estimation for the ISA portion, the off-line results computed earlier will be the "priors" for the Bayesian estimation using the Kalman filter algorithm.

Table 6.1 A summary of algorithm choices

Processing Stage	Preferred Algorithm	Other Algorithms	Comments
Off-line Block ML	Empirical Bayes Estimation	• Kalman Filter • Kalman Smoother	Smoother since future data is available off-line.
In-Stream Analytics **Dynamical ML**	Kalman Filter	More complex Markov Chains (many choices for the State Transition Matrix).	Exact Recursive solution; Use priors from the off-line stage.

SPECIAL COMBINATION OF THE BAYES FILTER AND NEURAL NETWORKS

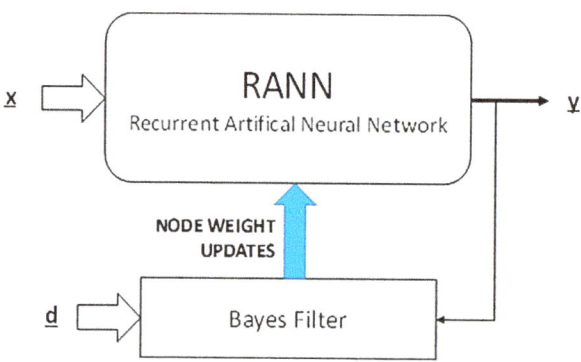

Figure 6.2 Combining the neural network and Bayes filter

Following the approach of Haykin [HS08] and others, we propose a general solution using a combination of the Recurrent Artificial Neural Network (RANN) and Bayes filter as (Figure 6.2). The Bayes filter algorithm are the Kalman filter and its variants. The Training Set is $\{x_i, d_i\}$. During testing, when there is no d_i, $\underline{y}$ is the ML output (classification or regression). During training, the RANN performs the forward calculations. The weight update is not performed in the RANN (for example, by back propagation).

The Bayes filter performs the weight updates and are passed up to the RANN for the forward calculations. The details of the RANN network topology are in Figure 6.3.

For the hidden layers, the following are true:

- Each layer has

    1. Network nodes – L in number
    2. Feedback nodes – L in number
    3. Bias node – 1 in number
    4. Total of $K = 2L + 1$ nodes per hidden layer

- Each network node performs a weighted sum of its inputs, which is passed through a non-linearity.

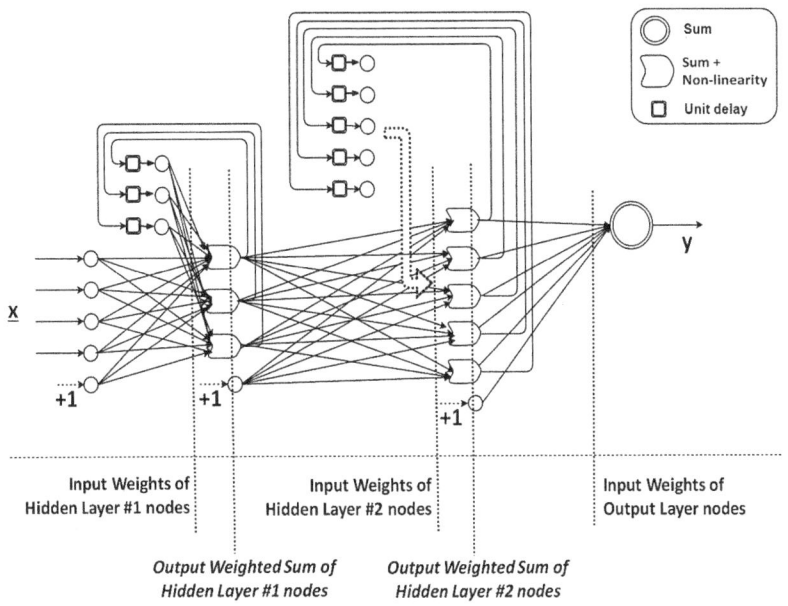

Figure 6.3 Neural network topology

- The feedback node delays the non-linear output of a network node by one unit.

- The bias node has a constant +1 as its input.

The output node is a linear summing node. The input node is a pass-through node.

Bayes Filter

We have some choices for the Bayes Filter algorithm.

1. Mild non-linear Gaussian case – Extended Kalman Filter (EKF).

2. Non-linear Gaussian case – Cubature Kalman Filter (CKF) or the Unscented Kalman Filter (UKF).

3. Non-linear distribution-free case – Particle Filter or the Markov Chain Monte Carlo (MCMC) Filter.

The EKF has many decades of successful practical use in widely varying applications. Therefore, we focus on the EKF algorithm, assuming that the non-linearity of our application is of the mild variety.

The RANN + BF is a supervised non-linear time-varying dynamical machine learning solution. We call this solution the *Bayes Recurrent Artificial Neural Network* (BRANN) algorithm. The generality of the BRANN algorithm should not be misinterpreted as simplicity. For the BRANN, the number of hidden layers, number of nodes, state transition matrix, and Bayes filter initialization parameters are all user choices. ML algorithms should have many features that allow us to adjust the solution.

The BRANN solution is the nice fit with our ISA implementation stack. The off-stream training of a RANN can be done using the BRANN. During the ISA operation, when "*d*" is available, the recursive update of the RANN weights (on-line learning) can occur using the Bayes Filter. If not, the RANN will operate with the most recent weights in the forward-propagation mode. Otherwise, the RANN is fully capable of learning the static mapping between input and output. Hence, the BRANN is an algorithm for in-stream and off-stream analytics.

REFERENCES

[YP11] Young, P, *Recursive Estimation and Time-Series Analysis*, Springer, 2011.

[MP96] Madhavan, PG and Williams, W, *Kalman Filtering and Time-Frequency Distribution of Random Signals*, SPIE Proceedings, 1996.

[SS13] Sarkka, S, *Bayesian Filtering and Smoothing*, Cambridge University Press, 2013.

[KT00] Kailath, T, *Linear Estimation*, Pearson, 2000.

[HS08] Haykin, S, *Neural Networks and Learning Machines*, Pearson, 2008.

THE KALMAN FILTER FOR ADAPTIVE MACHINE LEARNING

L et us consider the ISA application in more detail. During ISA operation, the supervisor for supervised learning (the quantity we refer to as the "desired response" of $d[n]$) is available at each time instant (or counter = n). It is this supervisor data, $d[n]$, that allows the algorithms we saw in the last chapter assess how much error is being made and how to adjust the weights so that the error will be less at the next instant, $[n + 1]$.

However, in practical situations, the supervisor, $d[n]$, may not be available from time to time. Then, our solution will have to operate with most recent weights updated when d was available and not do any more weight updates. Once $d[n]$ resumes, the weight update can also re-start. Hence, we need an algorithm for both in-stream and off-stream analytics. Let us sketch such a scenario in Figure 7.1.

Initially, in preparation to developing the solution, we will have massive amounts of data collected that can be used to train the solution in the off-line mode.

The Kalman filter has three modes of operation:

1. Predictor – before $d(n)$ is available at counte $r = n$

2. Filter – after $d(n)$ is available at counte $r = n$

3. Smoother – after all the training data is in hand

For the off-line case, use the Kalman Smoother. For the in-stream case, use the Kalman Predictor when $d(n)$ is not yet available (the Kalman filter updates after $d(n)$ arrives).

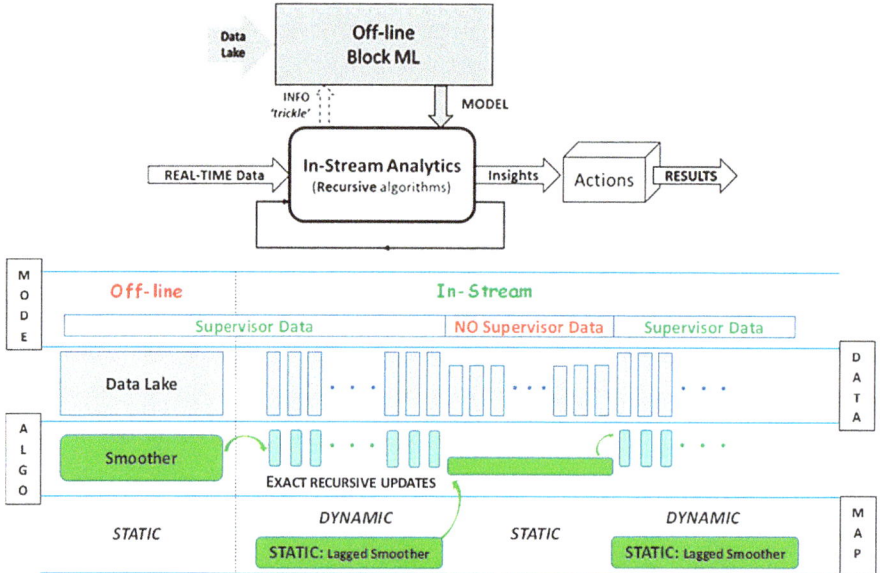

Figure 7.1 In-stream analytics operations

Instead of using the BRANN, we develop a new Kalman-filter-based solution for adaptive machine learning; we use the following elements:

a. Kernel projection machine learning (non-linearity)

b. Recurrence via output feedback (long term memory)

c. Time-Varying Linear Kalman filter (dynamics)

Towards the end of Chapter 4 (in Part I), we discussed kernel methods and random projection as two of the many options and coupled them with the recursive least squares. Here, we replace the RLS with the Kalman filter. As such, we call this solution the *Kernel Projection Kalman Filter* or KP-Kalman Filter which addresses all types of mapping we encounter in machine learning.

The KP-Kalman Filter is our recommended solution for adaptive machine learning for ISA.

KERNEL PROJECTION KALMAN FILTER

In Chapter 4, we discussed how random projection addresses the non-linearity of the separating surface among classes. The input to the Kalman filter is already transformed non-linearly by the random projection method. As RLS did in the case of RP-RLS in Chapter 4, here the Kalman filter performs a linear operation (finding the *linear* separating surface after the input has been transformed non-linearly into high dimensions, as per the cover theorem application). What the Kalman filter is called to do that RLS could not is to cope with the dynamics of the system, which the Kalman filter accommodates through the evolution of the state as a Markov process in the state equation of the state space model.

The architecture is as follows:

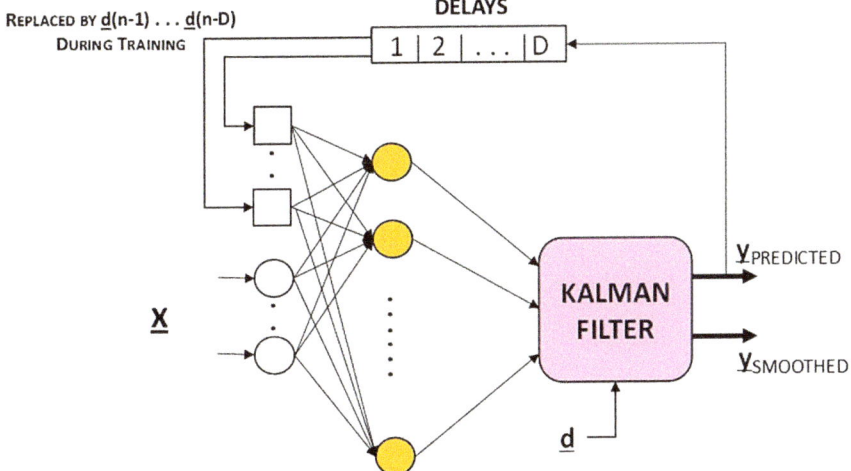

Figure 7.2 Kernel projection Kalman filter structure

Many careful architectural and parameter choices have to be made in the combined structure above. For the prototypical non-linear "double moon" classification problem [HS08], we make the choices as follows for the double moon dataset:

- Class 1 is red and Class 2 is green.

- As the black curve indicates, the separating surface is highly non-linear.

- Inputs are $\{x1, x2\}$, the x and y coordinates of each of the points on the "half moons." The desired responses, $\{d\}$, are the corresponding class memberships, $\{1, 2\}$. There are 2,000 data points in this plot.

- The dataset is presented to the KP-Kalman filter in a random order. Here, we select the randomization based on a simple Markov chain, which adds the dynamics. There is one probability for the next sample to stay on the same half moon and another probability for to jump to the other half moon.

■ *MATLAB "m" code for data generation is in the chapter appendix.*

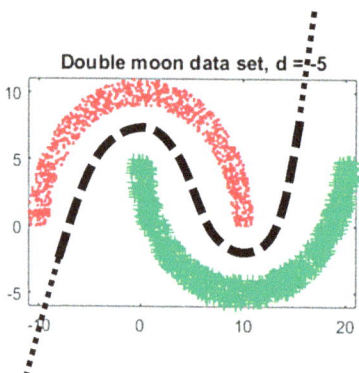

Figure 7.3 The double moon dataset and the kernel projection Kalman filter

A random projection has the following attributes in this example:

■ Gaussian non-linearity for all hidden nodes.

■ A sufficiently large number of hidden nodes (= 20 here).

■ Two delayed outputs were fed back to the input layer.

The Kalman filter has the following attributes in this example:

■ Three sets of subscripts for the output, states, and state covariance that are predicted, filtered, and smoothed.

For the state space model, the following are true:

$$\underline{s}[n] = \underline{A}\ \underline{s}[n-1] + \underline{q}[n-1]$$

$$y[n] = \underline{H}[n]\ \underline{s}[n] + \underline{r}[n]$$

where $\quad \underline{q}[n-1] = N[\underline{0}, \underline{Q}(n-1)]$ and $r[n] = N[0, R(n)]$ $\qquad$ (1)

where $\underline{A}$ can be constructed to handle more complex dynamics but here, $A = I$. A special choice of $\underline{H}[n]$ is made for the time-varying case. $\underline{H}[n] = [-\delta[n-1] \ldots -\delta[n-D]\ x_1[n] \ldots x_M[n]]$ where $\delta(n-D) = d(n-D)$ during training (called *teacher forcing*) and $y_p(n-D)$, the predicted output during the in-stream operation.

To understand the consequence of this choice, consider the delayed samples of $y[n]$.

$\underline{H}[n] = [-y([n-1] \cdots -y[n-D] \, x_1[n] \ldots x_M[n]]$. With this choice of $\underline{H}[n]$, we can write the following from Equation (6):

$$y[n] = -s_1[n] \, y[n-1] - \cdots - s_D[n] \, y[n-D] + s_{D+1}[n] \, x_1[n] +$$
$$s_{D+M}[n] \, x_M[n-M+1] + r[n].$$

(This is the well-known *transfer function model* developed by Box and Jenkins in 1976.)

In our case of the KP-Kalman filter, $\underline{H}[n]$ contains the non-linearly transformed inputs and delayed outputs created by the random projection structure. As shown in Figure 7.4, the Kalman filter estimates the weights for each output of random projection network $[= \rho(s_m[n])$ for $n = 1$ to $N]$ recursively.

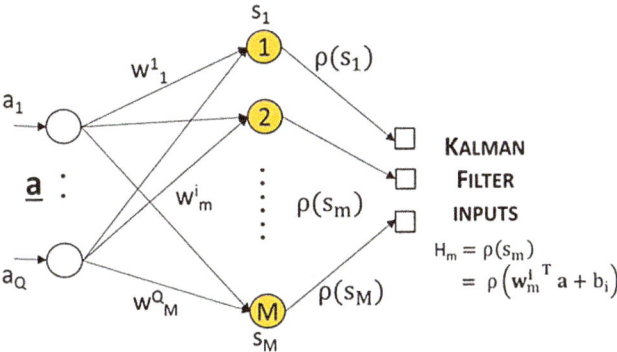

Figure 7.4 Input to the Kalman filter

In our case, $\underline{a}[n] = [-\delta[n-1] \cdots -\delta[n-D] \, x_1[n] \ldots x_M[n]]$.

Summary for linking Kernel Projection to Kalman Filter

Choose M and the function $\rho(.)$.

Assign random numbers to all $\underline{\mathbf{w}}_m^i$ and b_m.

At each n, the Kalman filter exogenous input $= [\rho(s_1) \cdots (s_m) \cdots (s_M)]^T$.

Subscripts: *P – Predicted estimate; F – Filtered estimate; S – Smoothed estimate.*

Initialization: $s_F[0] = \mathbf{0}$; $P_F[0] = l^{-1} I$ with λ = small positive constant; $Q[0] = Q = I$; $r[n] = r = 1$; $A = I$.

In the algorithm updates below, the notation for the states has been revised. The "s" in Equation (1) and the "s" with subscripts below are different.

The states with subscripts, s(.) = conditional expectation, E[s|x, d]. which is the Bayes estimates obtained from the conditional probability density function, p[s|x, d] (without explicitly estimating the pdfs).

Counter, n [=1 to N]:

$$\underline{H}[n] = [\rho(s_1[n]]) \cdots \rho(s_m[n]]) \cdots \rho(s_M[n])]$$

Prediction:

$$s_p[n] = As_F[n-1]$$

$$P_p[n] = AP_F[n-1]A^T + Q$$

$$\rightarrow y_p[n] = H[n]\,s_p[n]$$

Filtering:

$$v[n] = y[n] - H[n]\,s_p[n-1]$$

$$M[n] = H[n]\,P_p[n]H^T[n] + r$$

$$K[n] = P_p[n]\,H^T[n]\,M^{-1}[n]$$

$$s_F[n] = s_p[n] + K[n]\,v[n]$$

$$P_F[n] = P_p[n] - K[n]\,M[n]\,K^T[n]$$

$$\rightarrow y_F[n] = H[n]\,s_F[n]$$

Smoothing occurs once the previous updates are completed.

Initialization: $s_S[N] = s_F[N]$; $P_S[N] = P_f[N]$

Counter, n [N – 1 to 1]:

$$s_p[n+1] = A\,s_F[n]$$

$$P_p[n+1] = A\,P_F[n]\,A^T + Q$$

$$G[n] = P_F[n]\,A^T\,(P_p[n+1])^{-1}$$

$$s_S[n] = s_F[n] + G[n]\,\{s_S[n+1] - s_p[n+1]\}$$

$$P_S[n] = P_F[n] + G[n]\,\{P_S[n+1] - P_p[n+1]\}\,G^T[n]$$

$$\rightarrow y_S[n] = H[n]\,s_S[n]$$

This completes the Kalman filter + Smoother algorithm. The MATLAB m code for the KP-Kalman filter + Smoother is in the chapter appendix.

Kernel Projection Kalman Filter *Double Moon* Experiment

There are two stages in this experiment, as shown in Figure 7.5.

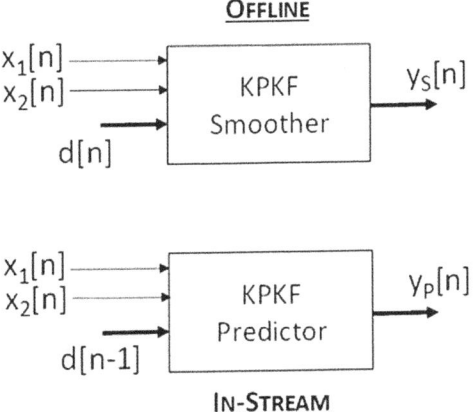

Figure 7.5 Off-line and in-stream operations

1. Offline

- All Training Set data is used.

- At counter $= n$, both $\{x_i[n], d[n]\}$ are available.

- The output of interest is the *smoothed* Kalman output, $y_S[n]$.

- We will also inspect the *smoothed* states, $\underline{s}_S[n]$, which are the conditional expectations, $E[\underline{s} \mid x, d]$.

- $y_S[n]$ is obtained from the conditional expectation of states via $y_S[n] = \underline{H}[n] \underline{s}_S[n]$ where $\underline{H}[n]$ are known and hence non-random quantities.

  $\therefore y_S = E[y \mid x]$, the conditional expectation.

2. In-Stream

- At the start of counter $= n$, $\{x_i[n], d[n-1]\}$ are available.

- The output of interest is the *predicted* Kalman output, $y_P[n]$.

- The *predicted* states, $\underline{s}_P[n]$, are the conditional expectations, $E[\underline{s} \mid x, d]$.

  $y_P[n] = \underline{H}[n] \underline{s}_P[n]$.

- Once $d[n]$ arrives, the Kalman filter updates $y_F[n]$ for the next recursion.

To make the figures clear, the plots in Figure 7.6 show a brief 200-sample Training Set experiment.

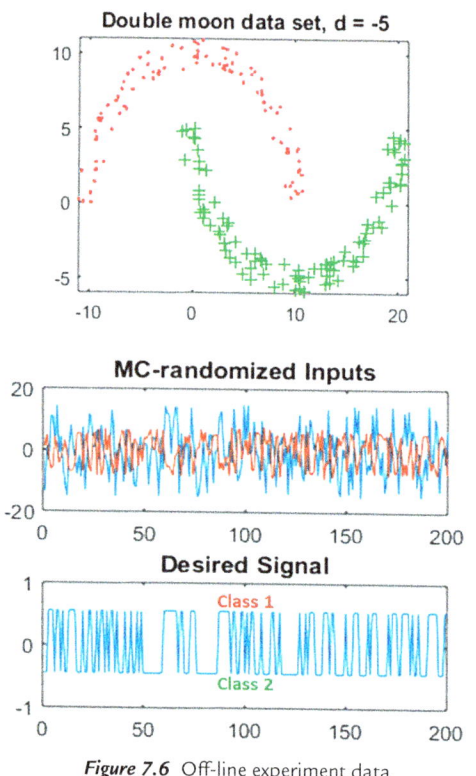

Figure 7.6 Off-line experiment data

Offline: Large amount of $\{x_i, d_i\}$ typically; only $i = 1$ to 200 in this experiment.

- The double moon data is randomized using the Markov chain. On the right, $\{x_i, d_i\}$, $i = 1$ to 200, are shown on top and bottom, respectively.

- Supervised learning is performed using the Kalman Smoother.

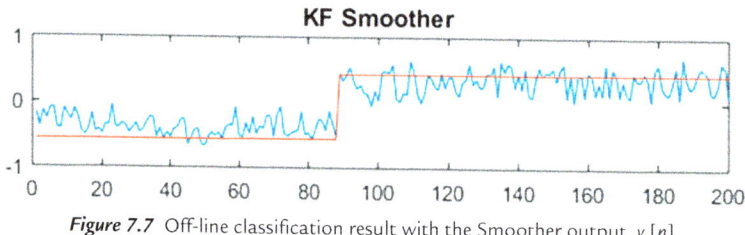

Figure 7.7 Off-line classification result with the Smoother output, $y_s[n]$.

Note that the output of the KP-Kalman filter is in random order. It has been re-ordered in Figure7.7. For classification purposes, the threshold was selected as "0" and the continuous output was discretized into Class 1 and Class 2.

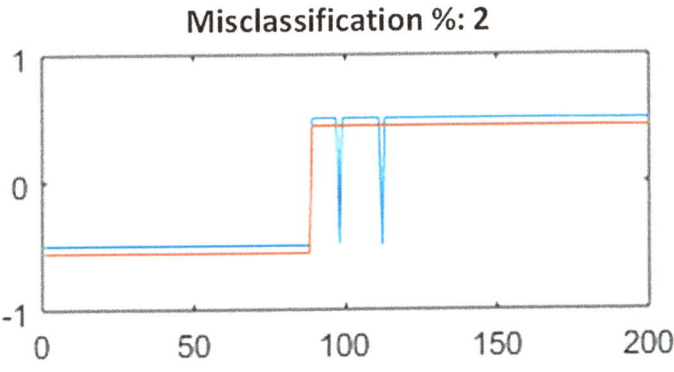

Figure 7.8 Classification results for off-line

The misclassification error is only 2%. The performance of the KP-Kalman filter is excellent for a highly non-linear and dynamical classification problem.

Similar results were seen on multiple re-runs. The results were more stable in a run longer than just 200 data points. We also note that the smoothed states (which are the weights of the output layer of the random projection network) were stabilized.

As we have noted, the Kalman Smoother requires "future" data for its backwards operation. Hence, this version of the KP-Kalman filter is to be used during the off-line training phase. Once the converged and smoothed state vector is obtained, one can simply use it in the in-stream portion as the fixed weights of the output node of the random projection network. Another preferred way for the in-stream portion is discussed in the next portion of the experiment.

Experiments conducted with the RLS as in the RP-RLS method in Chapter 4 produce inferior results to the KP-Kalman filter due to the dynamics, as expected. As a baseline, the linear least squares solution (using the pseudo-inverse) was also performed, which produced very poor results due to the non-linearity of the problem at hand (as expected).

For the in-stream classification, we have the following:

- At counter $= n$, assume that the inputs, $\{x_i[n]\}$ have arrived, but not $d(n)$ yet.

- Past values of $d(n)$ are available, therefore, we used Teacher Forcing.

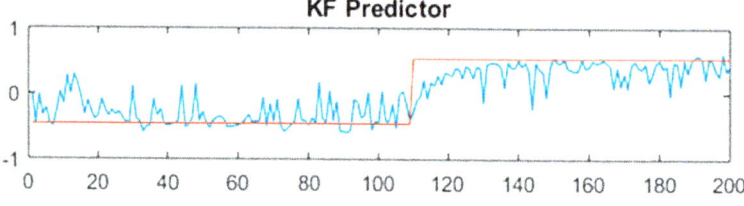

Figure 7.9 In-stream classification results showing the predictor output, $y_p[n]$.

As one would expect, the predictor shows the convergence behavior noticeable near 0 and 100. However, note that this plot is already sorted. In actuality, they are simply the first few data points from each class that the Markov chain randomization had produced and presented to the KP-Kalman filter.

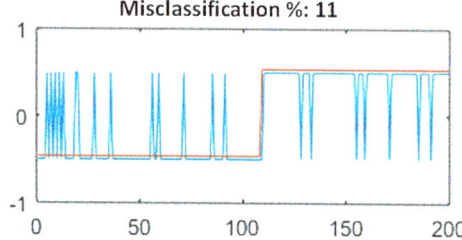

Figure 7.10 Classification result for the in-stream approach

The misclassification is larger (11%) in Figure 7.10, as we are working with the predicted quantities in the Kalman Filter. The plot of the state vector in Figure 7.11 makes it obvious that the convergence is not completed in 200 points. As can be expected, for a much longer data stream, the states (or weights) would have converged more and the misclassifications would have eventually disappeared. On longer runs, that is exactly what we see.

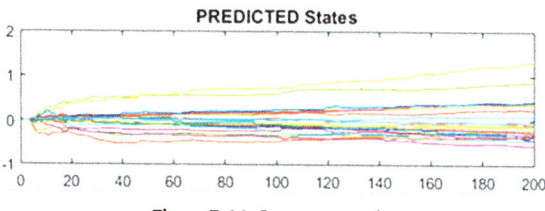

Figure 7.11 State vector plot

However, it should be noted that we expect the in-stream predictor solution to perform worse than the fff-line Smoother solution since the smoother uses more data (from the future).

Once $d(n)$ for the current counter, n, has arrived, the rest of the Kalman filter and Smoother updates can the performed, ready to repeat the in-stream cycle for $(n + 1)^{th}$ instant. The recurrence (in this experiment, $D = 2$) has a negligible effect on the Smoother and Predictor State estimates. Since the Markov chain-driven double moon data does not have any long-term memory, we do not expect the two delayed samples of output being fed back to the input stage to have any effect.

As we see, the KP-Kalman filter is a complete solution for in-stream processing. We start with stored data and smoothed estimates of the states (or weights). As each new data point arrives, the KP-Kalman filter is updated and the classification is predicted using the Kalman $y_p[n]$ output. Recursively, the weights are updated at the end of each period, n. This approach is clearly superior to using converged weights (=states) from the off-line Smoother (equivalent to the long-term prediction) for all of the in-Steam processing duration.

OPTIMIZED OPERATION OF THE KP-KALMAN FILTER

Some additional improvement should be possible during the in-stream operation. You can experiment on your own, but consider the following approach:

- The Kalman Smoother comes in multiple varieties. We have implemented the Fixed Interval Smoother. Another one of interest is the Fixed Lag Smoother.

- The Fixed Lag Smoother uses data up to some past the counter limit to smooth the estimates that came before it.

- It is conceivable that if the appropriate Fixed-lag Smoothed States can be used as the past state prediction vector ($\underline{sP}[n]$), the KP-Kalman filter predictions may be better. What one gives up is the "adaptation" or the ability to track fast changes in the dynamics of the system.

This MATLAB code is found on the companion files or by writing to the publisher at *info@mercuryleanring.com*.

REFERENCE

[HS08] Haykin, S, *Neural Networks and Learning Machines*, Pearson, 2008

THE NEED FOR DYNAMICAL MACHINE LEARNING: THE BAYESIAN EXACT RECURSIVE ESTIMATION

In the previous chapters, we saw the merger of the Bayesian exact recursive estimation (algorithm for which is the Kalman filter/Smoother in the linear, Gaussian case) and *machine learning*. We developed a solution called the Kernel Projection Recurrent Time-Varying (KPRTV) Kalman filter for business applications that require static or dynamical, dynamical or time-varying dynamical, linear or non-linear machine learning. Therefore, the KPRTV Kalman filter can be considered a "universal" solution.

Many university courses in ML largely teach static ML. Given a set of inputs and outputs, they train students to find a static map between the two during supervised "training" and use this static map for business purposes during "operation" (which is called "testing" during pre-operation evaluation). However, in real life, a static solution is not very useful.

We know that ML learns a "map" that relates the input and output of a system if the system does not change (i.e., it remains static). Static maps can be used for detection and prediction during the operational phase.

As an initial approach to practical ML, the static system assumption may be acceptable. However, we can go beyond such grossly simplified assumptions to develop more sophisticated solutions.

During the operation phase, detection involves noticing the changes in the ML map output when the underlying system undergoes changes. Prediction, however, involves quantifying the changes in the output as the system evolves. Clearly, detection is an easier task than prediction.

NEED FOR DYNAMICAL ML

Let's consider an IoT example for dynamical ML. Assume you are monitoring thousands of machines on a manufacturing plant floor with tens of thousands of monitoring points. In the training phase, "normal" operation for all these machines would have been established by deciding on a range of acceptable ML output values. Let us say multiple inputs from a machine (such as the vibration, temperature, and pressure) are trained against what is considered normal/abnormal. As we saw in the "Double Moon" experiment, the Kalman Smoother (and predictor) outputs will vary due to noise and other errors. We "threshold" this varying output to determine what is normal or abnormal.

When a bearing starts performing less than optimally on a machine, the ML output moves beyond the threshold and an alarm is generated. The appropriate action is then taken on the problem machine.

With a static ML solution, we can do this. However, what is normal for a machine (or the system under observation) is environment-dependent and ever-changing in real life.

Systems can operate in in a range in the normal zone, which we try to capture by pre-determining the threshold or range for ML output values. However, such a safe-range determination is ad-hoc because we do not have the actual experience of this particular machine and its own range of normal behavior as it ages. This lifetime evolution is one example of "wandering" in the normal zone. The consequence of the system's evolution within a normal range for static ML is the possibility of increased number of false positives.

When an abnormal situation is indicated, the prediction of the machine's condition (via the ML "map" output) will not be possible since the system has changed (as the detection of an abnormal operating situation indicated). In summary, static ML is adequate for one-off detection (and subsequent off-line intervention) if a business has a high tolerance for false positives.

In the dynamical ML solution case, the Kalman Smoother in the off-line training phase does everything that a static ML map can do. In addition, our Bayesian solution is optimal in the mean squared error sense.

STATES FOR DECISION MAKING

In any business application, a greater amount of relevant information is useful because it can be exploited to improve performance. Knowing that a machine has a problem is good, but knowing the nature of the problem is better. The Kalman predictor in the in-stream or operational phase provides the following:

- The prediction of the ML output – Where is the machine condition going beyond just a normal/abnormal indication provided by thresholding? Is the deviation large or small, indicating the severity of the problem?

- The predicted state trajectory. This is the true power of systems analytics. States are a sort of "meta"-level description of the machine. If we compare the predictor output and state trajectories in the previous sections, we can see that the states are less variable. This is an indication that the states are capturing better information about the underlying system.

At a simple level, if we move all the decision-making we did with the predictor output (and thresholding) to the state trajectories, the IoT solution performance will be better due to the less volatile nature of the states (there will be fewer false positives and false negatives). For example, consider the predictor output and state trajectory plots for the in-stream phase experiments in the previous sections and any one misclassification event (detected due to the prediction output exceeding the threshold). If you observe the state trajectories at the same instant of misclassification, there are hardly any changes. This indicates that this event is most likely a false positive and you do not have to send a technician to troubleshoot the machine.

Let us go beyond the use of System parameters for classification:

- With growing experience, changes in predictions of the output and states can be connected to the nature of change in the machine (both when normal and abnormal, providing a level of remote troubleshooting minimizing "truck rolls").

- This is an emerging domain-specific area that can have a significant impact on (1) predicting issues and fixing them before they result in an adverse business impact and (2) prescribing actions for improving business performance.

The New Normal

Let us say that a machine usually processes aluminum, but switched to titanium work pieces (the vibration and temperature signals are very different). The Kalman filter adapts to the new normal instead of retraining the static ML map. With static ML, one option would be (1) to train the static ML solution for multiple types of work pieces separately and switch the map when the work piece is switched (a tedious and error-prone solution) or (2) train the static ML map for all potential work piece types. This option damages the map and makes it less accurate overall; in probabilistic terms, this is caused by the non-homogeneity of data beyond heteroscedasticity.

Even though we have used the IoT as an example, it must be obvious that other business applications are close analogs.

If you subscribe to the view that true learning is a "generalization from past experience and the results of new actions," and therefore ML business solutions should be adjusted and re-applied on a regular basis, then every ML application is a case of dynamical machine learning. In the Kernel Projection Recurrent Time-Varying Kalman Filter, we have a Bayesian exact recursive estimation solution for dynamical machine learning that one can build on. The area is rich and there are many related algorithms that can be used for better results.

In summary, if your business problem space is static, use static machine learning. If not, use dynamical machine learning for the many practical benefits it brings. However, you must address the increased complexity and challenging underlying theory.

SUMMARY OF KALMAN FILTERING AND DYNAMICAL MACHINE LEARNING

Let's consider the key ideas we have developed so far in Part II of this book.

The purposes of machine learning (ML) in business applications are as follows:

1. Condition monitoring (How is my business doing today?)

2. Preventive maintenance (Predict issues and fix them before they have an adverse business impact.)

3. Performance improvement (Prescribe actions for customer acquisition and retention.)

When framed as the items in the prior list, all business applications of machine learning or data science are examples of in-stream analytics. In Chapter 5, we listed some examples of in-stream analytics applications.

- Fraud Detection

- Financial Markets Trading

- IoT and Capital Equipment Intensive Industries

- Health and Life Sciences

- Marketing Effectiveness

- Retail Optimization

Retail businesses that utilize ML solutions should understand that they must be adjusted and re-applied on a regular basis. In the applications listed above, the time constants may vary from milliseconds for fintech solutions to hours for health care solutions and weeks and months in retail approaches. But in all cases, closed loop updates are necessary. Therefore, every ML application can be a case of in-stream analytics.

We developed two classes of in-stream analytics solutions by merging the Bayes theorem and cover theorem applications. First, we discussed BRANN (Bayes Recurrent Artificial Neural Network). The main feature was that the Bayes filter is used to update the weights of the artificial neural network.

The other approach is the Kernel Projection Kalman filter ("KP Kalman filter"). Kalman Filter operates as the output layer of the ML method for non-linear mapping that implements the cover theorem. This approach controls the "curse of dimensionality," since the number of Kalman states is not tied to the number of weights of the artificial neural network.

A diagram that combines the two classes is shown in Figure 8.1.

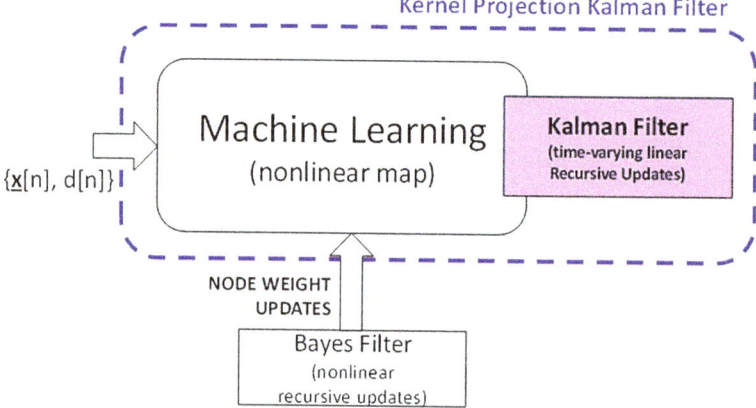

Figure 8.1 Combination of the BRANN and KP Kalman filter

Solution Summary

In-Stream Solution	Algorithm	Recurrence
BRANN	Bayes Filter	Yes
KP-Kalman	Kalman Filter	Yes

Bayes Filter: Required to be non-linear (dynamics via recurrence)

1. Mild Non-linear Gaussian case – Extended Kalman Filter (EKF).

2. Non-linear Gaussian case – Cubature Kalman Filter (CKF), Unscented Kalman Filter (UKF).

3. Non-linear distribution-free case – Particle Filter, Markov Chain Monte Carlo (MCMC) Filter.

KP-Kalman Filter:

- Linear filter

- Accommodates time-varying dynamics via the Kalman filter state equations and recurrence

---

The KP Kalman filter is a good choice for all machine learning business applications.

The KP-Kalman filter can be used for applications that require static or dynamical, dynamical or time-varying dynamical, linear or non-linear mapping.

Machine learning involves adjustments. The KP-Kalman filter is no exception. Here are some additional notes for this approach:

Architectural choices: Number of Kernels (hidden nodes); Number of delayed outputs for recurrence

Parameter choices: Kalman filter initialization values

Earlier, we discussed an approach to improve performance using a fixed-lag Kalman Smoother in a judicious manner. The Kalman filter applications to machine learning are indeed numerous.

CHAPTER 9

DIGITAL TWINS

D igital twins are an important part of the IoT, especially in Industrial IoT. They allow us to visualize the motor or pump on a machine and manipulate its "avatar." This allows us to accomplish many important goals:

- See the current condition of the equipment

- Compare the current data to the stored past data

- Compare the machine to similar machines

- Understand failures and improvement possibilities

- Try out changes in software and examine the possible effects of interventions

- Assess gains and risks

- Aggregate the digital twins together to assess system-wide performance

There is a progression of the digital twin versions mentioned in this book that deliver increasing levels of value.

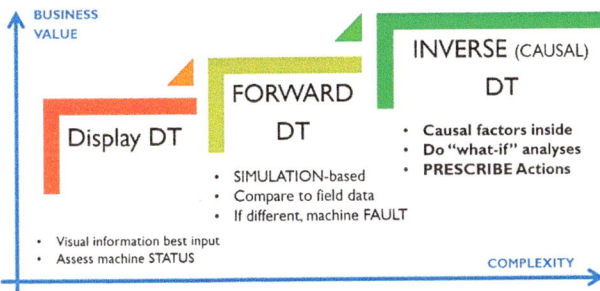

Figure 9.1 Types of digital twins

Digital twins can display data, can be used in software simulations, and support inverse causal modeling (ML+ systems theory).

Let us briefly consider each of the important aspects of digital twins.

1. Displaying data

Seeing field measurements on your PC displayed on a rendering of the machine is revelatory. You can view all the data overlayed on a picture of the machine and apply either visual or ML-assisted fault detection.

2. Software simulation

Once we have a "companion" software simulation of the machine, we can compare the field measurements with the simulated "measurements." Any deviation is an indication of an issue – either that the simulation is inaccurate (and needs updating) or more importantly, something is wrong with the machine.

Since you have a complete simulation in hand, you can investigate further. What changes to the simulation will match the field measurements? Are these changes within the simulation's limits? If not, you have a problem in the field which requires a "truck-roll" and root-cause analysis in the field. Such a process saves money and time by avoiding disrupted operations and the needless deployment of expert troubleshooters who are often in short supply.

3. Inverse CAUSAL models

When you have deployed Inverse DT, you have gone beyond the field measurement level. The Inverse DT exposes the underlying system that generated the measurements that you displayed.

The measurements by themselves are not important. There is nothing controversial in this statement. When we measure the vibration signals from near a bearing, it is not the vibration itself that is of interest. We want to know if the inner or outer race or balls have problems.

The real purpose of developing and deploying a digital twin is to understand what the parameters of the system are so that we can develop a CAUSAL understanding of what the measurements are telling us.

What Inverse DT does is to systematically incorporate a software simulation and add machine learning and system identification (for estimating the underlying parameters) on top. This is the power of Inverse DT: being able to incorporate an independent source of information generated by the simulation into traditional estimation methods.

CAUSALITY

All of today's ML is correlation-based. We know that correlation is NOT causation! A recent article, "How Causal Inference Can Lead to Real Intelligence in Machines" [SR19], does a good job of explaining this predicament . Most of the credit for the "causality crusade" goes to Judea Pearl (the Turing Prize winner) who started his investigations into causality calculus starting in early 2000.

There is a new awakening in AI and ML that incorporating causality is essential to move the field forward. This is because while association or correlation-based ML has brought us many benefits, it is not sufficient to get us the valuable insights that lead to prescribing specific actions to achieve specific results.

When you take a medicine, you know that randomized controlled trials have been conducted to prove the cause-effects of that medicine. This is the gold standard method to prove causality. A physician will not prescribe the medicine if it was only correlated with your illness. In the same way, causality is required for prescriptive analytics.

You cannot perform randomized controlled trials for industrial machinery. Causality is challenging, and requires you to employ multiple strategies to accomplish your goals. We are at that juncture now with IoT and digital twins.

A digital twin encompasses AI, machine learning, and data science when the input is from the IoT. The cause-effect determination of what is happening with an object occurs in the digital twin. This is the role of the inverse digital twin (IDT).

INVERSE DIGITAL TWIN

The concept of Inverse (causal) DT twin can be best explained by a medical scenario. Let's assume you go to a neurologist since you are feeling unwell. She will ask you to get an EEG (electroencephalography) study where the technician will place multiple electrodes on your head and measure and display the electrical activity (a form of display DT of your brain). The specialist can make certain conclusions based on past experience, such as "spindles" will appear in the measured data when sleeping (which is an example of correlation studies done in the past). The neurologist may want to know the origin of the epilepsy signature that had appeared in the measured EEG. So, from measured data on the surface, she can deduce the cause of the signature. This is an example of inverse modeling. Friston and his colleagues [FK19] worked on this problem of the dynamic causal modeling of the brain.

EEG data are measured on the scalp, but the specialist wants to know the epileptic locus inside the brain of a live human, which is not easy to do. Going from the activity inside the brain to the EEG on the surface is the forward problem and the more challenging task of identifying what is happening inside the brain from the EEG on the scalp is the inverse problem. This example clarifies a distinction that has bedeviled the digital twin community: what is the real difference between the simulation DT and causal DT? To avoid this confusion, we use the terms forward and inverse digital twin, respectively.

In the case of an IoT digital twin, the Forward DT is a software simulation, such as NPSS or PCB Heatmap. Forward DT has a strong history and is used by numerous companies. However, the Inverse DT has not been developed (yet). That is because the inverse problem is an "ill-posed" problem in mathematics and is considered impossible to solve generally (other than by using multiple constraints).

Finding the causes given the effects (surface measurements) is difficult, but rewarding. We know the real causes of issues in an operating machine and we can prescribe the right actions to solve them at the root level.

In mathematics, the constraint that is usually used to solve an ill-posed problem is called *regularization* (of Tikhanov). This approach involves choosing the solution vector with the smallest norm (in L2 regularization). However, such a solution may have no bearing on reality while still being mathematically satisfying. We need solutions that will lead directly to a useful outcome, such as identifying the location of a bent shaft inside a motor.

INVERSE MODEL FRAMEWORK

Application-domain constraints make the solution to the ill-posed problem meaningful. Consider the IoT's industrial machinery applications. Let us start with the general framework shown in Figure 9.2.

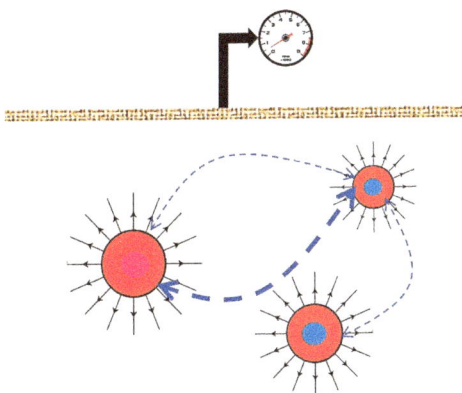

Figure 9.2 General framework for inverse modeling

The meter indicates a surface measurement device (like the EEG machine). Below the surface (in brown), there is a "generator" that is a collection of three signal sources of varying strength and location. Based on the radiation of the signals from these three sources, the meter will measure the additive combination of these three sources in the simplest case.

Now assume that these three sources are also "coupled" (via wired connections perhaps). As the coupling strength changes, so do the dynamics of the system and the signal measured by the meter. Coupling changes can give rise to very complex measured surface data from which deducing the number, location, and coupling dynamics of the sources can seem virtually impossible. This is why we call this an "ill-posed" problem.

In the rest of this discussion, we assume that our main interest is in the dynamics of the system. In the specific case of machine design, we are interested in the kinetics and kinematics of the machine and not the statics or structural aspects. (There are DT applications where the static and structural aspects are important. One case involves the design of a product using CAD/CAM techniques, where we can visualize the stress and thermal distribution and improve the design.)

Consider a real-life example of a motor to be monitored. Vibration is usually measured on the surface (body) of the motor. Considering the dynamics of the motor that we want to track, the two bearings at either end and the shaft connecting them are the key sources of vibration when the motor is operational. Much like the previous fictional meter example in Figure 9.2, vibration from these parts travel to the body and combine to provide surface measurements.

When the inner or outer race of the left or right bearing is faulty, the vibrations will increase. But this change will affect the shaft also. If the shaft is misaligned or slightly bent, both bearings will be affected and vibration from all three components may increase. The interactions can be considered as "couplings." As these couplings change dynamically over the life of the motor, the vibration patterns will also change. Note that such couplings are sometimes called *virtual sensors*.

GRAPH CAUSAL MODEL

Let us start using the traditional causality approach. At the risk of over-simplifying, here is a way an IoT engineer can understand the basic tenets of causality.

The focus here is on the structural aspects and not the sensor data time series generated by each asset. Time dependencies will be discussed later.

Consider four entities (assets), B1, B2, B3, and B4, as shown in Figure 9.3. The questions to ask are as follows: (1) Do all the links exist? (Some entities may not be connected); (2) If so, what are the directions of the links?; and (3) What is the strength or weight of the links? The first two questions are related to the causal discovery (of the structure) and the third question is related to the causal estimation (statistical/ signal processing methods when the data are sensor-measured time series from each of the four assets).

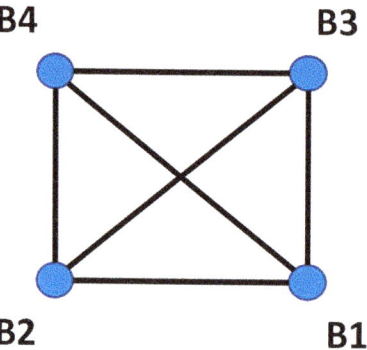

Figure 9.3 Example of the possible relationships between entities in a system

At the risk of oversimplifying this example, these are some of the main concepts in causal structure discovery.

- A causal model is *sufficient* if it does not contain unobserved common causes or latent variables.

- *Markov assumption*: For causally sufficient sets of variables, all variables are independent of their non-descendants in the causal graph conditional on their direct causes (parents in the causal graph).

- *Faithfulness*: Causal influence is not hidden by the coincidental cancellations.

- In directed acyclic graphs (DAGs), one can use d-separation ("d" for "directional") to identify the pair-wise independent nodes, which means that there is no link connecting the pair. *d-separation* is a mechanical procedure that answers the equivalent conditional independence question.

These factors have to be satisfied for causal discovery and estimation to be valid, which is very challenging to prove in traditional applications in the social sciences. For example, if the nodes of the graph are Health, # cigarettes smoked, Age, and Obesity, one can imagine how difficult it will be to draw the correct directional links of the DAG while asserting the Markov and faithfulness assumptions.

CAUSALITY INSIGHTS

Classification and regression and causal inference are different, according to Peter Spirtes, who is a long-time proponent of causality at CMU; he discussed the probabilistic reasoning in one of his recent works [SP16]. There are important differences between the problem of predicting the value of a variable in an unmanipulated population from a sample (classification and regression) and the problem of predicting the post-manipulation value of a variable from a sample from an unmanipulated population (causality); the latter is called *counterfactual analysis*, which reveals the difference.

Causal factors in a DAG are pair-wise regression coefficients when the Markov and faithfulness conditions are satisfied and d-separation is applied to the graph. The direction of the link (regress X on Y or Y on X) can be addressed by estimating the regression coefficient in each case, generating the residuals, and applying statistical test for independence. If X truly causes Y, the residuals for the regression in this direction will be independent. Regression in reverse will generate residuals that are uncorrelated (by definition) but not independent.

Once given a DAG with the conditionally independent links between the nodes removed by d-separation and the direction of the link determined by checking the residuals, we have the *causal graph*. In practical use cases, we will have an outcome variable of interest; for example, in a market research study on a soap brand, the reported customer satisfaction or acceptable price may be the outcome variable and other nodes may be color, smell, and size. The market research team will distribute questionnaires to many people and collect these data, which form the sample data at each node.

From the market research study data collection example, it must be clear that time is not a variable in this causal analysis. This is called the structural model or instantaneous causality. Consider the soap brand marketing study. The measurements are from different consumers at different times, but the time-stamp has no bearing on the analysis. Such structural causality is the realm of traditional causality. In a general case, though, the question arises of delayed or "lagged" measurements and their causal dependence.

In the IoT case, the factors influencing the outcome variable will have temporal dependencies in addition to the structural ones. Since IoT measurements are typically time-series-measured from multiple assets interconnected (such as on a plant floor), instantaneous and lagged causality are important. If the locations of each asset in the layout are significant, the structural model is spatially distributed. In the most general case, IoT data for DAG is a spatio-temporal multichannel time series.

The next step in the development of the Inverse DT is the creation of the Graph Causal Model. This is where domain experts come in who can tell us how best to form the model. For example, in the motor example, a mechanic who has spent a lifetime repairing and rebuilding motors can tell you the key sources of vibration and how the sources are causally related.

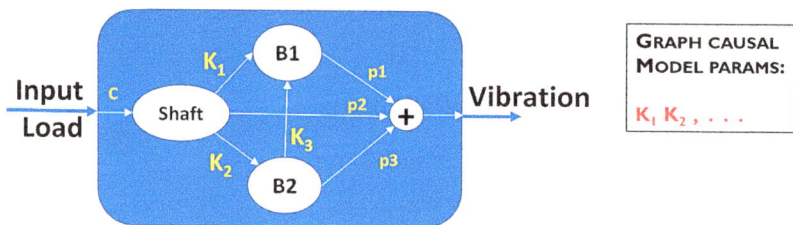

Figure 9.4 Graph causal model of motor vibrations

The resulting overall model of the motor for our dynamics analysis may be as shown in Figure 9.3. The load input, "c," can be measured as the current draw or by a torque meter. K_1, K_2, and K_3 are the couplings of interest. p_1, p_2, and p_3 are the less important (for our study of motor dynamics) transmission strengths of vibrations to the sensor on the body of the motor.

This Graph Causal Model is the forward simulation model for our Inverse DT solution and involves the interplay between the pp. forward and inverse models within our Inverse DT algorithm.

INVERSE DIGITAL TWIN ALGORITHM

To get familiar with Inverse DT architecture, consider a plant whose digital twin is of interest to your business. The plant is a known entity because either we made it (say, a motor) or we have a model with known parameters, $\{p\}$, and field measurements (say, the soil moisture and nitrogen from a wheat field). Our purpose with the Inverse DT is to track the variations in $\{p\}$ in real-time from the field measurements. We also consider a special case when a field measurement is not available, such as when a sensor failed or that measurement is practically impossible to make (perhaps due to cost).

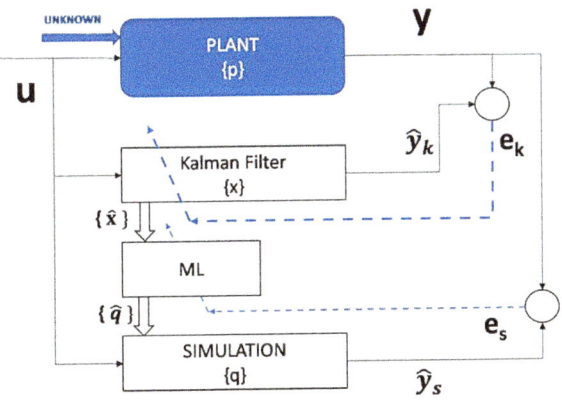

Figure 9.5 Inverse Digital Twin algorithm structure

The traditional role of the Kalman filter is to act as the *observer*, which reveals the plant's parameters. With just the input and output of the plant available to a Kalman Filter, the estimated states are not, in general, estimates of $\{p\}$ when certain system matrices are unknown. The added blocks of machine learning and software simulation of the plant in Figure 9.6 are important to recovering the plant parameters. After sufficient learning, $\{q\} \approx \{p\}$.

The challenging part of Inverse DT algorithms is the development of the combination of a neural network and the software simulation, Ψ.

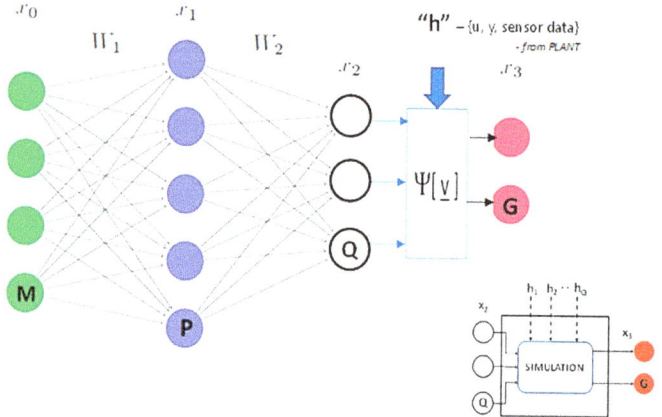

Figure 9.6 Combining neural network and simulation

The MIMO neural network is shown in Figure 9.6 with one hidden layer was implemented using a matrix version of the back propagation algorithm. The M inputs are the M states from the Kalman filter. The values of $\{x_2\}$ are the $\{q\}$ software simulation parameters, which when converged are the estimates of the plant's parameters, $\{p\}$.

In the first case, assume that the plant output y and all the field measurements are available. The software simulation, Ψ, uses these inputs and the running simulation produces estimates of the plant output y and the other field measurements. The differences between the actual and those produced by Ψ gives us the MIMO error vector. Once you have the error vector, one has to back propagate the error across Y and through the neural network layers all the way to the neural network MIMO input layer. All of them are standard work, except the back propagation of the error vector across the software simulation, Ψ.

The propagation of the error vector back across Ψ and the later stages of the neural network requires a re-derivation of the algorithm (nothing more than applying the Chain Rule carefully).

In the second case, we deprive ourselves of one of the field measurements for running Y and show that the plant parameters can still be estimated adequately. Clearly, one would expect that as more field measurements are unavailable, you will be unable to estimate $\{p\}$. Even if you have a software simulation, you cannot save the full cost of sensors and networks for field measurements.

SIMULATION

To explain our simulation test of the algorithm, we use a simpler case than the motor example discussed earlier. Let us consider an agricultural crop model shown in Figure 9.7, where the crop yield is the output. We start with the genetic potential of the seeds and properties of the crops that can be measured, such as the seed weight. This stochastic model provides an estimate of the crop yield that can be expected, C.

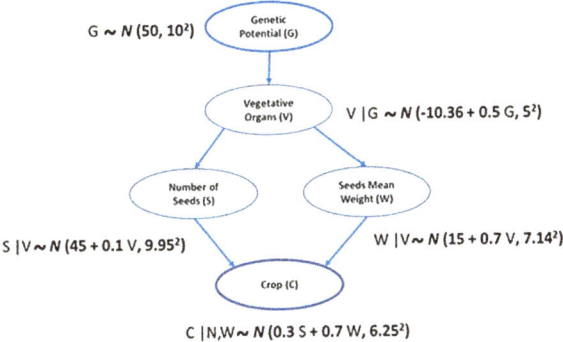

Figure 9.7 Causal model of an agricultural crop

Case (ia): Crop Field Test with All the Measurements

Here are the equations of the plant and the plant model used to "create" the plant in the crop model (Figure 9.8).

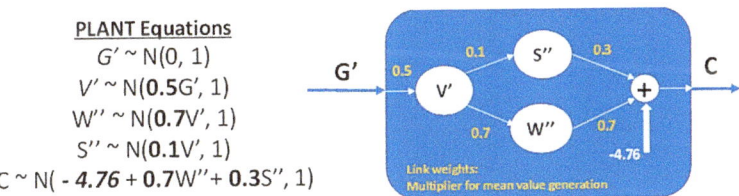

PLANT Equations
$$G' \sim N(0, 1)$$
$$V' \sim N(\mathbf{0.5}G', 1)$$
$$W'' \sim N(\mathbf{0.7}V', 1)$$
$$S'' \sim N(\mathbf{0.1}V', 1)$$
$$C \sim N(\mathbf{-4.76} + \mathbf{0.7}W'' + \mathbf{0.3}S'', 1)$$

Figure 9.8 Simplified crop model

A few observations are in order.

- The variable names and the equations are from an agricultural crop example.

- G' is a stochastic process with a Normal distribution (with zero-mean and unit-variance) that is not dependent on any antecedents.

- All the subsequent stochastic processes have the previous stochastic process as the mean.

- The graph model has a bifurcation and modes not visible from the output.

A realization of all the variables (the field measurements) is shown in Figure 9.9. A close inspection of the time series traces show the effect of the means being random; they are not simply scaled values of each other. In a non-real-time case, if we took each pair and performed linear regression between them, you would recover $\{p\} = \{0.3, 0.7, -4.67, 0.7, 0.1, 0.5\}$, to some level of accuracy. However, remember that our objective is to estimate them "on the fly" iteratively so that we can track their variations over time. This is the key to Inverse DT so that we can do a "what-if" analysis.

The bottom section of Figure 9.9 shows the Kalman filter states. In general, they are not estimates of $\{p\}$.

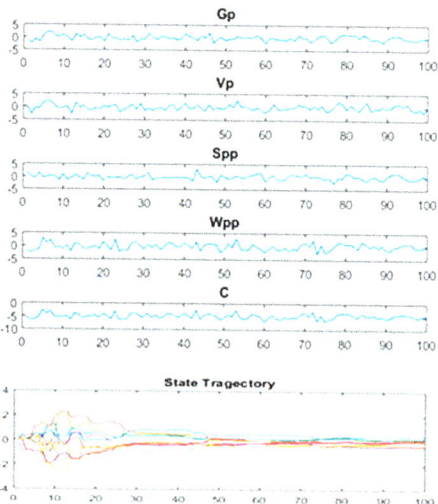

Figure 9.9 Typical realization of each of the graph variables and Kalman states

With these states as inputs to the MIMO neural network with the software simulation, Y, embedded in it, we get the trajectories of $\{q\}$. After convergence, the estimated $\{q\}$ is very close to actual $\{p\}$ - below.

The estimated parameters are

$$0.3000 \quad 0.7000 \quad -4.7600 \quad 0.6366 \quad 0.0748 \quad 0.5551$$

The true parameters are

$$\mathbf{0.3} \quad \mathbf{0.7} \quad \mathbf{-4.67} \quad \mathbf{0.7} \quad \mathbf{0.1} \quad \mathbf{0.5}$$

We can observe the trajectories of $\{q\}$ in Figure 9.10. The slowest trajectory (yellow) is that of the bias term, minus 4.67.

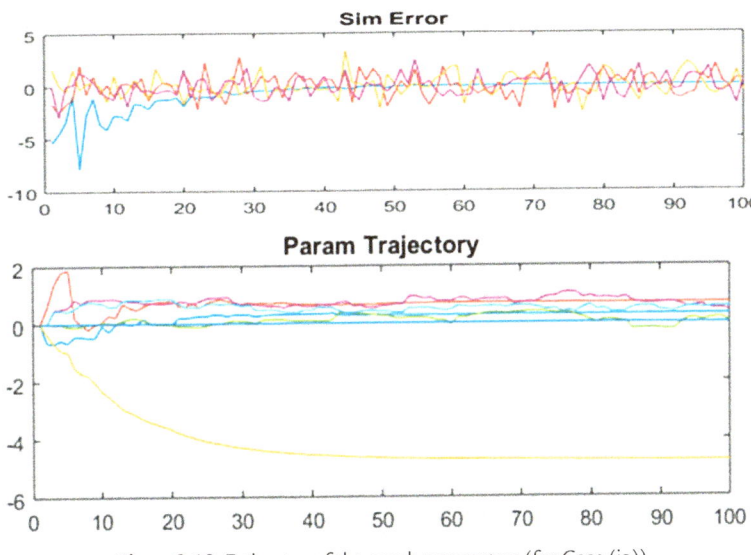

Figure 9.10 Estimates of the graph parameters (for Case (ia))

Case (ib): Crop Field Test with All measurements and Tracking of the Parameter Shift

This case is identical to the previous example, except that at $n = 50$; the parameter between Vp and Spp jump-shifts from 0.1 to 1.0.

The $Vp - Spp$ parameter estimate is isolated in the lower portion of Figure 9.11. Within the random variations, one can clearly see the estimated tracking of the variation from 0.1 to 1.0 with some learning time delay.

The estimated parameters at $n = 100$ are

$$0.2999 \quad 0.7001 \quad -4.7558 \quad 0.6822 \quad 1.0310 \quad 0.4871$$

The true parameters at $n = 100$ are

$$\mathbf{0.3} \quad \mathbf{0.7} \quad \mathbf{-4.67} \quad \mathbf{0.7} \quad \mathbf{1.0} \quad \mathbf{0.5}$$

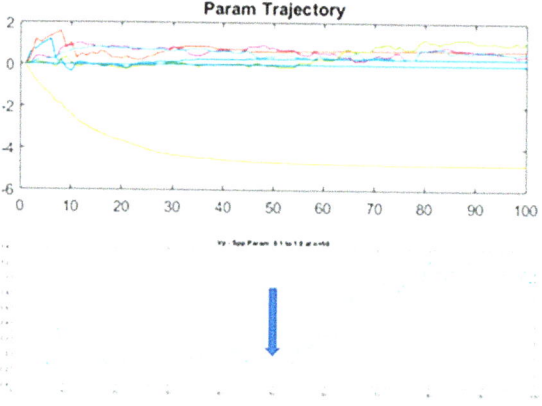

Figure 9.11 Estimates of the graph parameters for Case (ib)

Case (ii): Crop Field Test with One Measurement Missing and Tracking of the Parameter Shift

This is the case when a field measurement is not available, such as when a sensor failed or that measurement is practically impossible to make.

In this demonstration, we assume that Wpp is not available to Inverse DT algorithms; this means that the software simulation, Ψ, does not receive the Wpp measurement as indicated.

Remember that we have the graph model and parameters from when we first developed the simulation of the crop field. This is a great advantage; within Ψ, we can create the *Wpp* measurement that we do not have as follows:

$$Wpp = 0.7 * (\text{field-measured } Vp).$$

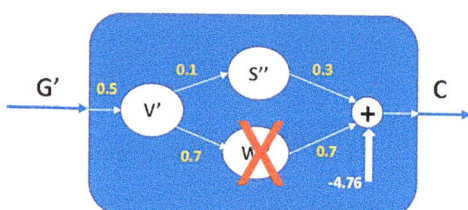

Figure 9.12 Crop field test with one measurement missing

The work that went into creating the simulation model and initial parameters is consider information. This information can be cleverly exploited in the estimation process.

Figure 9.13 shows the successful estimation of all other parameters, including the jump in the $Vp - Spp$ parameter, which was 0.1 and changed to 1.0 at $n = 50$.

The estimated parameters at $n = 100$ are

0.2715   0.7082   −4.7560   xxx   0.9463   0.4724

The true parameters at $n = 100$ are

0.3   0.7   −4.67   0.7   1.0   0.5

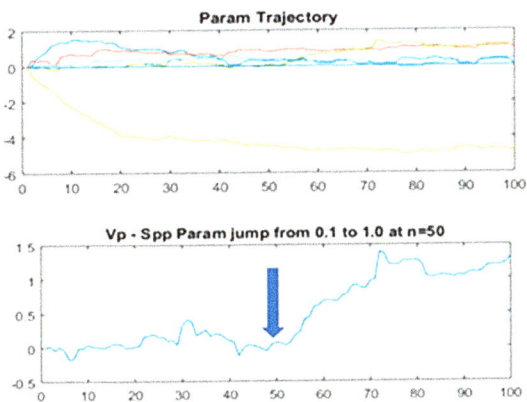

Figure 9.13 Estimates of the graph parameters with the missing and jumping parameters

CONCLUSION

We have introduced a new class of digital twins called the Inverse DT. The name emphasizes the fact that Inverse DTs provide the facility to alter the digital twin on our PC, answer the "what-if" questions and even try out counterfactual scenarios so that we can provide valuable prescriptive analytics to business owners.

The Inverse DT is a simulation-based digital twin that learns the simulation parameters of a specific machine and tracks the variations in the system parameters over that machine's operating life.

The agricultural crop model that inspired the tests has a non-trivial graph structure and doubly-stochastic data models. We introduced parameter jumps and showed that the Inverse DT is able to learn and track parameter variations. From a signal processing point of view, the Kalman filter in this application is a systematic way of improving the signal-to-noise ratio of the MIMO neural network inputs. The states of the Kalman filter do not provide the plant parameter estimates directly unless the measurement matrix is known.

The plant's differential equations cannot be constructed easily (other than in some man-made situations, such as airplane frame dynamics). Even in "natural" plants, such as those in an agricultural crop field, it may be possible to develop the system of equations from the diffusion and evaporation physics for specific limited cases. However, in general, we will need the additional MIMO neural network with the software simulation embedded in it.

We see the algorithms of Inverse DT as sufficient to track parameter variations, as well as offer some work-around when some sensors fail. This is the benefit of the Inverse DT, which allows us to incorporate an independent source of information generated by the simulation into traditional estimation methods. We are not manufacturing information from nothing. We are exploiting the knowledge (and the information) contained in the simulation structure and some guesses of the governing parameters to enhance "on the fly" real-time estimation for tracking the plant (or other things, such as machines or crop fields).

The truly exciting aspect of Inverse DT is the interplay between the software simulation and parameter estimation methods. There are many ways to construct the simulation, Ψ, which will lead to clever ways to infuse valuable a-priori information into the estimation step. There is no single general solution. Data scientists can use their ingenuity and creativity to develop many combinations of simulation-estimation solutions in partnership with domain experts.

REFERENCES

[SR19] Sagar, R, *How Causal Inference Can Lead to Real Intelligence in Machines*, Analytics India Magazine, 2019.

[FK19] Friston, K, et al., *Dynamic causal modelling revisited*, Journal of Neuro-image, 2019.

[SP16] Spirtes, P and Zhang, K, *Causal discovery and inference: concepts and recent methodological advances*, Applied Informatics, 2016.

A NEW RANDOM FIELD THEORY

We are introduced a new development stochastic process theory which may be valuable for random processes in higher dimensions than one (which is typically a time series). This theory and its practice will apply to two, three, and higher dimensional random fields.

Random field theory is the generalization of the stochastic process or time series theories to multiple dimensions. In undergraduate courses on the topic, students deal with one-dimensional random fields, almost always the time dimension. Remember, a stochastic time series is a set of random variables ordered over time. An excellent generalization of such random processes with some remarkable results was fully developed from scratch and published as a book by Erik Vanmarcke in 1983 [VE83] when he was a professor at MIT.

Vanmarcke introduced a comprehensive theory of random fields which extends to multidimensional cases. His random field theory of local averages captures the effect that local averaging has on a homogeneous random field. The quantification of the effects of local averaging leads to a function that characterizes the second-order properties of the random field called the *variance function*. A scalar called the *scale of fluctuation* derived from the variance function has many interpretations, some historic and some new. The scale of fluctuation can be considered to be similar to other scalars derived from multi-dimensional probability density functions such as correlation or Shannon entropy. The information that the scale of fluctuation provides is different from others; in the case of a time series, the scale of fluctuation is a measure of the time scale of a random process over which the correlation structure of that random process is characterized.

For the full exposition of the random field theory, refer to Vanmarcke [VE83]. (My work on this topic involved extensions to Linear Time-invariant

Systems and Kalman filter-based estimation method [MP97]).) We will proceed by mentioning some highlights and the potential applications in the current data science and machine learning fields.

Vanmarcke showed that the power at zero frequency is called the *Scale of Fluctuation* or q in random field theory.

$$\theta = \pi g(0) = \int_{-\infty}^{+\infty} \rho(\tau)d\tau$$

The equation and Figure E.1 are meaningful. In the figure, the height, (g(0) times π), and the area under the normalized autocorrelation function, $\rho(\tau)$, marked in blue are θ. This is the graphical meaning of Vanmarcke's equation for θ.

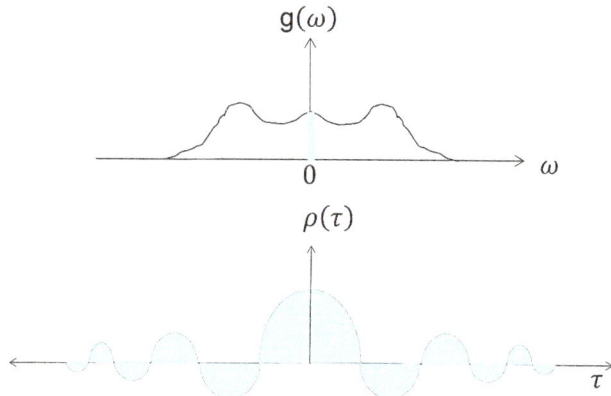

Figure E.1 Normalized spectral density and autocorrelation functions

Many interpretations for q exist and are variously known as the characteristic length, correlation length, or duration of persistence of trends. For a real-valued, discrete-time autoregressive random process, $X(n)$ of order P, we can obtain the scale of fluctuation, θ, as follows.

$$\theta_x = \theta_w \frac{\sigma_w^2}{\sigma_x^2} |H(0)|^2$$

$$= \frac{|H(0)|^2}{\sigma_x^2} \text{ where } |H(0)|^2 = \left[\frac{1}{1 + \sum_{i=1}^{P} a_i} \right]^2$$

Here, $H(w)$ is the transfer function of the linear-time invariant system relating the white noise process to the AR(P) process, θ_w is the scale of fluctuation of the white noise process $(\theta_w = 1)$, σ_W^2 is the variance of the white noise process $(\sigma_W^2 = 1)$, σ_X^2 is the variance of the AR(P) process and $\{a_i\}$, $i = 1$ to P, are the AR coefficients.

The results shown earlier are for a time series. For the 2-D case, θ is $4\pi^2 g(0,0)$ of its 2-D normalized spectral density (the same pattern follows for higher dimensional random fields). The calculations of θ for the 1-D and 2-D cases are straightforward (the ways to calculate the instantaneous values of θ using Kalman filtering are available in my past publication, *Instantaneous Scale of Fluctuation Using Kalman-TFD & Applications in Machine Tool Monitoring*). Some curious properties of θ for second order linear time invariant systems were also developed there.

Let us recap the property highlights and motivate future applications of θ in data science. From discrete-time linear time-invariant system principles, we know that the constant damping ratio and undamped natural frequency contours in the z-plane for a second order system are as shown on the left in Figure E.2.

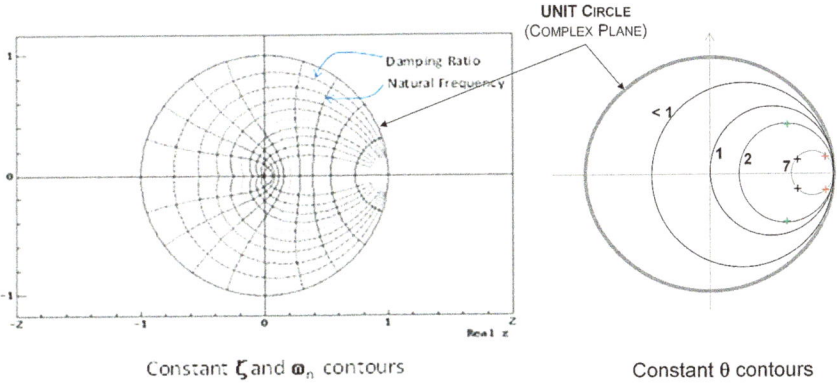

Figure E.2 *Constant q contour for a second order system*

It is notable that for a second-order system, the constant θ contours shown on the right have remarkably simple geometric shapes. In fact, for $\theta = 1$, the equation is quartic, but very similar to a circle with origin at $(0.5 + j0)$ and radius = 0.5. The equation for $\theta = 1$ contour is $(x^2 + y^2)^2 + x^2 + y^2 - 2x = 0$.

There is some similarity to the contours of the constant damping ratio, but there are also distinct differences. Similar to Shannon's scalar, entropy = H, and θ reduces the joint probability density function to a scalar.

Does θ capture some aspect of reality that is useful? The constant θ contours seem to imply significance as fundamental as a natural frequency and damping, but at this time, further insights are not obvious.

Some real-world applications of θ (see its use for the machine tool chatter prediction) point to the following physical insights.

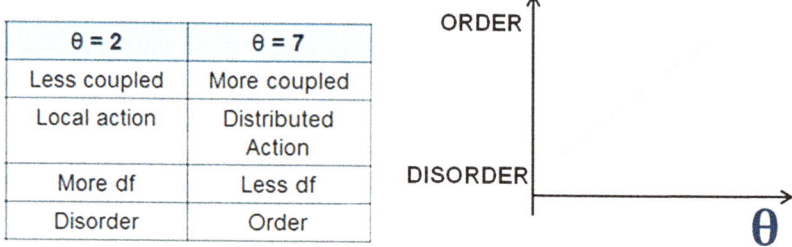

$\theta = 2$	$\theta = 7$
Less coupled	More coupled
Local action	Distributed Action
More df	Less df
Disorder	Order

Figure E.3 Physical properties of θ

While highly speculative, previous studies suggest that θ is proportional to coupling and to order in a distributed node system, whereas it is inversely related to the degrees of freedom (df), as indicated in Figure E.3. The concept is that more degrees of freedom result in a challenge for a distributed system where different parts are de-synchronized and they can spin off in different directions(pandemonium can ensue).

The hope is that the physical model of network activity utilizing the concept of explicit or implicit deep structures with internal coupling will help advance our analytics tools for the extraction of patterns and information from spatially and temporally distributed networked systems.

Let us reconsider our discussion from Chapter 9 about causal digital twins. We described a motor and its rotating elements, such as bearings and spindle, and explored how a causal chain of events propagate through the machine via the couplings K_1, K_2, and K_3.

As mentioned in the previous page, θ, or the Scale of Fluctuation, may also be indicative of different levels of coupling. As such, monitoring instantaneous values of θ of the IoT signal from the motor may be an early indicator of a degenerate coupling that can lead to malfunction of the motor. In the machine tool chatter case, that was exactly the indication that low θ precedes chatter. With this knowledge, a control scheme can be developed using θ such that the machine tool is operated at maximum capacity (highest safe speed and feed) to maximize production but minimize waste (which is created once the chatter phase is entered).

It is also noteworthy that there are some open questions for which there are no obvious responses. As such, they are mentioned here for the reader and future researchers in the area to pursue (and graphically shown in Figure E.4).

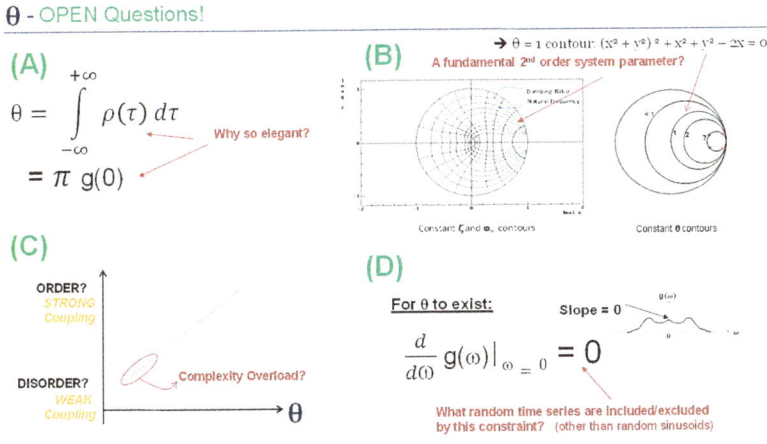

Figure E.4 Open questions about θ

a. θ has a strikingly compact definition. Why?

b. As briefly discussed above, the constant θ contours above seem to imply significance as fundamental as natural frequency and damping. Does θ capture some aspect of reality that is useful?

c. Is θ a true measure of complexity? There are many types of complex systems; which one does θ capture? Is weak coupling in a distributed system indicative of impending disaster in all cases?

d. In Vanmarcke's derivation of θ, there is only one constraint, that the slope of the spectral density at zero must be zero. Given the diversity of the spectral density, does this constraint exclude any major classes of random processes?

Some highly speculative use cases for θ are as follows:

■ In an electricity distribution system, monitor θ as an indicator of an imminent brown out.

■ In a flexible manufacturing system, monitor θ as a potential indicator of non-normality; drive θ "to the edge" such that production with a high quality is maximized.

■ From the many vital signs monitored for a patient, calculate the value for θ that may indicate the need to transfer the patient to the Intensive Care Unit.

There are as many use cases as you can imagine!

REFERENCES

[VE83] Vanmarcke, E, *Random Fields: Analysis and Synthesis*, MIT Press, 1983.

[MP97] Madhavan, PG, *Instantaneous Scale of Fluctuation Using Kalman-TFD and Applications in Machine Tool Monitoring*, SPIE Proceedings, 1997.

INDEX